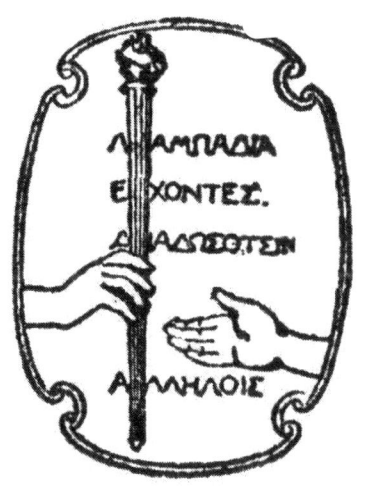

Round the year with the stars

stars

The chief beauties of the starry heavens as seen with the naked eye

Garrett P. Serviss

Originally Published by
Harper & Brothers Publishers
1910

Contents

PREFACE   1

INTRODUCTION   3

I THE EVENING SKY AT THE VERNAL EQUINOX   9

II THE EVENING SKY AT THE SUMMER SOLSTICE   23

III THE EVENING SKY AT THE AUTUMNAL EQUINOX   33
    FOOTNOTES: 43

IV THE EVENING SKY AT THE WINTER SOLSTICE   45
    FOOTNOTES: 56

V THE PLANETS   57

APPENDIX URANOGRAPHY OR HEAVENLY DESCRIP-
    TION OF THE CHURCHMEN   63
    LETTERS OF THE GREEK ALPHABET EMPLOYED IN
        URANOGRAPHY   65

PRONUNCIATION OF STAR AND CONSTELLATION
    NAMES   73

INDEX   77

i

PREFACE

This book represents an attempt to cultivate the love of the stars, and to offer a guiding hand to all those who are willing to believe that some of the most exquisite joys of life are to be found, like scattered and unregarded gems, waiting to be picked up by any chance wayfarer who, without special knowledge, or optical aids, or mathematical attainments, or any of the paraphernalia or advantages of the professional astronomer, will simply turn his eyes to the sky and open his mind to its plain teachings and its supernal inspirations.

The writer's only real excuse for appearing again in this particular field is that he has never yet finished a book, and seen it go forth, without feeling that he had overlooked, or cast aside, or of necessity omitted a multitude of things quite as interesting and important as any he had touched upon. Accordingly, he yields once more to the lure of this inexhaustible and illimitable subject, and strives again to find expression for the thoughts which it continually awakens, and to exhibit, however imperfectly, the endless procession of marvels which stream before him who knows and loves the stars like a dazzling *rivière* of brilliants.

This book in no way duplicates another work of the same hand, *Astronomy with the Naked Eye*. In *that* the effort was to revive the romance of the constellations by retelling their fascinating history, their mythology, their immemorial legends and traditions, and indicating their poetic background in the presence of the imaginary figures which, "from times of which the memory of man runneth not to the contrary," have been associated with them; in *this* the writer tries to draw the reader into more intimate relations with the stars by dwelling upon their individual peculiarities and beauties, and the impressions which either singly or in constellated groups they make upon the mind of the beholder. Surely there is not another field of human contemplation so wondrously rich as astronomy! It is so easy to reach, so responsive to every mood, so stimulating, uplifting, abstracting, and infinitely consoling. Everybody may not be a chemist, a geologist, a mathematician, but everybody may be and ought to be, in a modest, personal way, an astronomer, for star-gazing is a great medicine of the soul. There is the writer's text.

INTRODUCTION

The charts illustrating this book have been drawn by the writer especially to meet the needs of beginners—of those who, feeling what a void in their intellectual life ignorance of the stars has created, would now fill that void, and thus round out their spiritual being with some knowledge of Nature on her most majestic and yet most beautiful and winning side.

On account of the necessarily diminutive scale of the charts, everything has been omitted from them which did not seem essential. But for the purpose in view they gain by this process of exclusion, for with more details they would have been confusing. It is the broad, general aspect of the sky with which the beginner must first familiarize himself. At the start the heavens appear to him to be filled with an innumerable multitude of scintillating sparks, scattered everywhere in disorder. But with a little attention he perceives that there is discipline in this host, and immediately the discovery gives him pleasure and awakens his admiration, as the perception of order always does. The great leaders of the firmament come forth, unmistakable, plainly recognizable, and thereupon the rank and file fall into their places. Then the ineffable beauty of the whole assemblage bursts like a revelation upon the mind. This revelation is not for the dull in spirit, but he who has once had it becomes henceforth, and even in spite of previous prejudice or indifference, a devotee of the stars, with a zeal flaming brighter with every swing of the pendulum of his years.

In the four circular charts representing the aspect of the heavens respectively at the Vernal Equinox, the Summer Solstice, the Autumnal Equinox, and the Winter Solstice, few stars fainter than the fourth magnitude are included, and not all even of that magnitude, because the sole purpose is to enable the beginner to recognize the constellations by their characteristic groupings of stars and their relative situations in the sky. The insuperable difficulty is to picture the *hemispherical* sky on a *flat* page. A certain amount of distortion cannot be avoided, and the reader's imagination must supply the effect of perspective. He must always remember that the centre of the chart stands for the middle of the sky *overhead*, and that the circular boundary represents the full round of the horizon, from east through south, west, and north, to east again. If he is comparing the chart with the sky while facing south, he should hold the chart upright as it is printed in the book; if he makes the comparison while facing north, he should turn the chart upside down. If he lies on his back with his head to the north (and in no other way can one get so vast an impression of the starry dome), and holds the chart over his head, it will represent the entire vault of the firmament.

The names of the constellations will be found on the charts, and also the individual names of the most celebrated stars, but the constellation boundaries are not shown, because, in nine cases out of ten, the precise limits of a constellation are not important for the beginner to know, and to search for them would simply lead to confusion. As he progresses in his knowledge of the sky any uncertainty about the constellation to which particular stars belong can be settled by consulting the six charts, drawn to a larger scale, at the end of the book. On *these* charts more of the small stars are shown, and in addition there will be found the Greek letters which astronomers attach to the principal stars of each constellation for the sake of ready identification. On these charts, too, the constellation boundaries will be seen, indicated by dotted lines. The tracing of these lines is more or less a matter of arbitrary choice. There are no international boundary disputes among the heavenly powers, and the frontier lines may run anywhere, provided only that they do not include in one constellation any stars which by common usage, or prescription, belong to another. The constellations have been reshaped many times in the past. The "geography of the heavens" has known as many changes as that of the earth, the ambition of the old astronomers being sometimes as insatiable as that of founders of terrestrial kingdoms and empires. About three centuries ago the starry sky was "Christianized," St. Matthew, St. Peter, St. John, St. Joseph, St. Michael, St. Stephen, St. Gabriel, St. Mary Magdalen, St. Katharine, together with Noah, Aaron, Job, and Eve, taking the places of the heathen gods, goddesses, and heroes in the sky, while Saturn became Adam, Jupiter Moses, Mars Joshua, Mercury Elias, Venus St. John Baptist (!), the Moon the Virgin Mary, and the Sun Christ (see Appendix). It is not an unheard-of thing in uranography ("description of the heavens"; analogue to geography) for a star, or a group of stars, to change allegiance, or even to belong to two constellations at the same time. The bright star Alpheratz is still an example of this double nationality, for, although it shines on the head of Andromeda and is her jewel *parexcellence*, yet her neighbor Pegasus also lays claim to the star, and uranographers so far admit the justice of his claim that they call Alpheratz, according to circumstances, either Alpha ( ) Andromedæ or Delta ( ) Pegasi.

For many of their purposes astronomers find no use for the constellations, preferring to identify the stars by their position in right ascension and declination (equivalent to longitude and latitude), and in the great modern *Durchmusterungs*, or star catalogues, this plan is universally followed. Still, the constellations afford a very convenient classification of the stars, and probably they will never be abandoned even by professional astronomers; while from another standpoint they never can be abandoned, because they are among the most ancient and precious of human documents, valuable for history and for the understanding of mythology, and resistlessly charming in their poetic associations.

But, to return to the description of the charts, the reader should be informed as to the meaning of the lines shown upon them, and of the indications found round their borders. In the four circular charts the closed curve crossing the sky from right to left represents the equator of the heavens, which is directly over the equator of the earth; the vertical line through the centre shows the meridian, or north and south line, which, so to speak, follows the observer wherever he may go, occupying the same place in the sky, *at the same hour of localtime*, in all longitudes; and the dotted curve is the ecliptic, or the apparent annual path of the sun through the sky. The crossing points of the equator and the ecliptic are respectively the Vernal and the Autumnal Equinox, where the sun is at the two dates in the year when day and night are of equal length; and the farthest northern and southern points of the ecliptic are respectively the Summer and the Winter Solstice, where the sun is at the times of the longest and the shortest days in our hemisphere. These four fundamental points are all shown on the charts. Around the border the hours of right ascension are indicated by Roman numerals. Each hour corresponds to 15° of space, or one twenty-fourth of a circle of the sphere. The hours begin at the Vernal Equinox, which is graphically described as the "Greenwich of the Sky."

In the larger-scale charts at the end of the book the hours of right ascension are indicated at the bottom, and the degrees of north and south declination (the sign + standing for north and - for south) are shown at the side. In both cases the declination is reckoned from the equator. The four oblong charts of this series, taken together, represent the entire circuit of sky between 40° north and 40° south declination, and the two semicircular charts, taken together, show the stars within 50° of the north pole. Thus the entire set of six charts exhibits the complete dome of the heavens from the north pole to 40° south declination. In passing from the oblong to the semicircular charts it is only necessary to bring the hours of right ascension into accord. In the semicircular charts the hours will be found round the curved borders.

Each of the four circular charts in the body of the book represents the aspect of the *evening* sky at one of the equinoctial or solstitial epochs. To be more precise, these charts show the sky as it appears, at about the latitude of New York, at 10 P.M., on, respectively, March 20th (the Vernal Equinox), June 21st (the Summer Solstice), September 23d (the Autumnal Equinox), and December 22d (the Winter Solstice).

But the reader must not think that it is necessary to confine himself to the exact latitude, date, or hour just mentioned. Undoubtedly it would be better for the beginner to do that approximately, but it is not essential. The effect of a change of latitude is, perhaps, the least important. If the observer is farther south than about 40° north latitude, the southern stars will appear higher in the sky than they are shown in the charts, and some of the stars close to the northern horizon will sink from view. If, on the other hand, he is farther north (as in Canada or Northern Europe), the northern stars will appear higher, and some of those near the southern horizon will be invisible. But if he confines his attention to the stars and constellations represented in the central parts of the charts (which he should, in any case, do for other reasons), the effect of the shift due to difference of latitude will not be found very serious.

As to the effects of a departure from the hours and dates for which the charts are drawn, they, too, can readily be allowed for. Suppose that, without changing the date, the reader makes his observations an hour earlier than that given, say at 9 P.M., March 20th. Then he will find that some of the eastern stars, seen along the left-hand edge of the chart when it is held upright, have not yet come into view above the horizon, while other stars, not seen on the chart drawn for that date, are visible above the horizon in the west. To the stars thus carried out of, or brought into, view he should pay no attention; he will find them again on other charts when they are better placed for observation.

Next, suppose that without changing the hour of observation he changes the date, and instead of observing on the 20th of March he observes on the 5th. Then he will notice precisely the same difference that was manifest when his observation was made an hour too early on March 20th—*i. e.*, some of the eastern stars on the chart will not yet have risen, and other stars, not on that particular chart, will be visible in the west. Although at first all this may be a little confusing to the beginner, he will soon find that he can make due allowance for the changes of aspect. The whole matter becomes very simple when it is remembered that the heavens have a double revolution toward the west; one of these revolutions, due to the earth's rotation on its axis, being effected in twenty-four hours, and the other, due to the earth's revolution round the sun, requiring an entire year. One hour of the daily revolution (represented by an hour of right ascension) produces the same effect on the position of the stars as two weeks of the annual revolution; or two hours of the first correspond to one month of the second.

If the observations are made at a later date or a later hour than those indicated on the chart, the changes will occur in the reverse order—*i. e.*, western stars will have disappeared and eastern stars will have come up into view.

I particularly wish to impress upon the beginner the needlessness of being troubled about these discrepancies. He can avoid all possibility of perplexity by fitting his observations to the exact times of the charts. As I have already said, a difference of a few degrees in his latitude on the earth may be disregarded. The charts, with a slight allowance for the shift of position of the extreme northern and southern stars, are available for any of the middle latitudes of the northern hemisphere. And if the effects of a change of hour or date prove in the least confusing, the beginner has only to await the given date and the given hour, and all will be clear. Then, as soon as he has become familiar with a few of the leading constellations, the others, which in themselves are not so easily recognizable, will fall into their proper places, after which there can be no possibility of confusion. In fact, much less effort is required to become familiar with the aspect of the starry sky than is demanded for a similar acquaintance with the fundamental data of botany, mineralogy, geology, or any other of the observational branches of natural science.

It was at first the intention to indicate the course of the Milky Way on the circular charts by dotted outlines, but this was abandoned in view of the restricted space. Any one can easily trace the meanderings and branchings of this starry scarf, the contemplation of which carries the mind to greater heights of intellectual perspective than any other phenomenon of the world of matter. If the reader has the good-fortune to be situated where artificial lights do not interfere with the splendor of the heavens, he can observe the course of the Milky Way on any clear night; and, if he possesses skill in delineation, he may make charts of it and its offshoots which will be of real value. Better still if he has the means of photographing it. Here is a non-mathematical field of astronomy which is ripe for the harvest, and in which the laborers are few. The Milky Way is so full of wonders that centuries of observation and study cannot exhaust them. There is nothing more impressive than to see how it often follows curves of lucid stars as if some mysterious attraction were drawing it toward them; and yet it itself consists of stars.

A few more words of practical advice to the beginner. Let him, at first, confine himself to the bright and conspicuous stars and the striking groups shown in the charts, assigning each to its proper constellation. When he has become familiar with these in their broadest aspects, he can turn to the charts at the end of the book and familiarize himself with the constellation boundaries. After that, if he wishes to go further, as he almost certainly will, he can obtain a large star atlas, furnish himself with a telescope, and open up a new side of his life which will make him rejoice to be, for a few short years, a dweller on a planet inhabited by beings intelligent enough to lift their eyes above the horizon and to feed their minds with the inspirations of the universe.

Yet another thing, which may be a novelty to many, and which is sure to afford unexpected pleasure—when you have fairly learned the constellations, take a mirror and study them by reflection. This is a counsel of intimacy. Orion will seem less remote and more comprehensible when his living image is enclosed in a frame, which you can hold on your lap like an album. There is something startling in the sight of the starry heavens under your feet. I once enjoyed the sensation in perfection while stalking deer in a boat at midnight on the placid bosom of a forest pond. The water was as motionless as so many acres of black glass, and I forgot to look for the deer, in the shaft of light from the hooded "flare" at the bow, when we seemed to be drifting out into an ocean of ether, in the middle of the sky, with stars below as well as stars above. When we silently crossed the pond, and got far from the shores, the sensation was overpowering; it took one's breath away. We drifted right over the Milky Way, and Vega, Altair, and the "Northern Cross" gleamed beneath the keel. Be sure that your mirror is freshly silvered and clean, and remember the reversals of position which all reflections produce. If you hold the mirror before you inclined downward, the position of objects in the sky will be reversed top for bottom; if you hold it inclined upward, so as to see objects behind your head, they will be reversed right for left. With these precautions you will find the mirror a great convenience for studying constellations which are nearly overhead. It is the principle of the "diagonal prism" employed with telescopes, and of the hand-mirrors used by many visitors at the Vatican Palace to view with comfort the ceiling pictures of Michael Angelo in the Sistine Chapel. Thus the sky becomes an atlas, and you can study its living charts at leisure.

ROUND THE YEAR WITH THE STARS

I THE EVENING SKY AT THE VERNAL EQUINOX

The year has its morning no less unmistakable in its characteristic features than the dawn of the day. The earth and all of its inhabitants feel the subtle influences of the dawning year, and Nature awakes at their touch. This annual morning comes when the sun transits the equator, moving north, at the beginning of his long summer tour, about the 20th of March. This is the epoch of the Vernal Equinox, when the springs of life begin, once more, to flow. Then the sun truly rises on the northern hemisphere. Then the mighty world of the north, which Providence has made the chief abode of vital organisms on this planet, rouses itself and shakes off the apathy of winter, and men, animals, and plants, each after their manner, renew their activities, and in many cases their very existence. This annual reawakening is one of the profoundest phenomena of nature, and there are secrets in it which science has not yet penetrated.

Bliss Carman has beautifully pictured the terrestrial charms of the vernal season in his "Spring's Saraband":

"Over the hills of April,
With soft winds hand in hand,
Impassionate and dreamy eyed
Spring leads her saraband.
Her garments float and gather
And swirl along the plain,
Her headgear is the golden sun,
Her cloak the silver rain."

But why do not the poets see and express the hyperphysical charm of the spring evenings? When the light of the vernal day has faded the stars come forth, and in the quality of their shining reduplicate and heighten the impressions left by the quickening landscapes. More than half is lost if this be missed. But perhaps this side of nature is too transcendent even for poetry. One can behold but not tell it. Emerson has come nearest to its expression, and he puts it in prose:

"The grass grows, the buds burst, the meadow is spotted with fire and gold in the tint of the flowers. The air is full of birds, and sweet with the breath of the pine, the balm-of-Gilead, and the new hay. *Night brings no gloom to the heart with its welcome shade.* Through the transparent darkness the stars pour their almost spiritual rays. Man under them seems a young child, and his huge globe a toy. The cool night bathes the world as with a river, and prepares his eyes again for the crimson dawn."

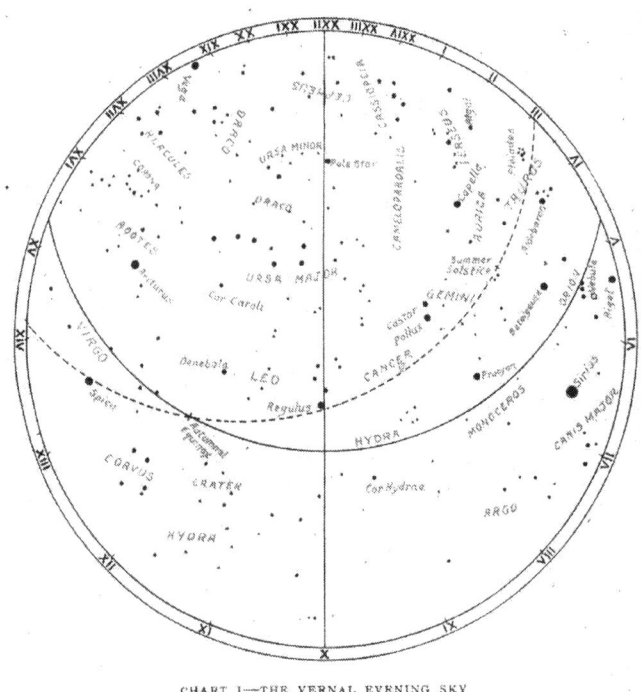

CHART I—THE VERNAL EVENING SKY

CHART I—THE VERNAL EVENING SKY

There was not only poetic but logical fitness in the old English custom, abandoned since 1751, of dating the opening of the year from the last week of March. How can the real birth of the year be imagined to occur when all nature is still deep in slumber under the January snows? The seasons are manifestly the children of the sun, waxing and waning with his strength, and surely that one should be reckoned the eldest which is the first birth of his vivific springtime rays. It seems remarkable that the beginning of the year in ancient times, when men felt more keenly than we do now the symbolism of natural phenomena, was not more frequently fixed at, or near, the Vernal Equinox, and I suspect some defect in our information on this subject. In Attica, George Cornewell Lewis tells us, the year began at the Summer Solstice. But this was to make the second of the sun's seasonal offspring the senior, thus ignoring the just claim of the true heir, the season of buds. In Sparta and Macedonia, according to the same authority, the year began with the Autumnal Equinox, which was still worse, for in summer the year is at the zenith of its life, while in autumn it is already stumbling toward the tomb. In Bœotia, at Delphi, and in Bithynia they contradicted nature more decidedly, as we do to-day, by making the year begin at the Winter Solstice, when the chilled world is yet asleep. The Romans adopted this plan eventually, but it is interesting to observe that they had an older custom of beginning the year in March, which many cherished in their domestic life as well as for some legal purposes, after the lawful opening of the year had been fixed on the 1st of January. And finally *we* have perpetuated the illogical system of absolutely reversing nature's rule in the succession of the seasons by making the year begin about nine days after the Winter Solstice. But in spite of calendars and laws nature prevails, and everybody instinctively feels that the true beginning of the year is in the season when the currents of life resume their youthful flow. At any rate, however it may be with strictly terrestrial affairs, that is the time when the call of the stars becomes the most insistent and irresistible. Accordingly the epoch of the Vernal Equinox is chosen for our opening chapter. But the reader at the commencement of his star-gazing is not confined to this season; he can begin at any time convenient to him.

To avoid misapprehension it is important to point out that our concern is not with that half of the heavens which the sun illumines when he crosses the equator, coming north, at the Vernal Equinox, but with the diametrically opposite half, where in countless multitudes shine his fellow suns—his peers, his inferiors, and his superiors—turning physical night into intellectual day. Accordingly, in Chart I we see not that part of the sky which contains the point called the Vernal Equinox, but the opposite part, where the sun pursues his course when he is declining from the Summer Solstice toward the Autumnal Equinox. The chart represents the appearance of the sky at 10 P.M. on the 20th of March (see Introduction). It also represents the sky as it appears about 11.30 P.M. at the beginning of March, about 9 P.M. the first week of April, and 8 P.M. about April 20th.

Let us, then, near one of these dates and hours, go out-of-doors and transport ourselves to the universe. Why does not everybody feel the thrill that comes to the astronomer when, with eager expectation, he watches the fading sunset light, the slow withdrawal of the vast curtain of illuminated air which for twelve hours has hidden the prodigious marvel of the spangled heavens, and the first peering forth of the great stars? I believe that everybody *does* feel it when he gives himself the opportunity and abandons his mind to its own reflections—but so few embrace the opportunity or encourage the reflections!

Select, if possible, a high place, where the eyes can range round the whole horizon unobstructed. Then try to seize the entire view at once, as one glances for the first time at the map of a new country. Get the *ensemble* by sweeping all around the sky, not pausing to note details, but catching at a glance the location of the brighter stars and those that form striking groups. Note where the Milky Way runs, a faint, silvery zone at this season, arched across the western half of the firmament, hanging like starry gossamers in places, brilliant in the northwest, but becoming fainter as it dips toward the southwestern horizon—a mere anticipation of its summer splendor, hiding its light and fading away as it approaches the imperial presence of Sirius. Notice the great hexagon of first magnitude stars that surrounds Orion in the west—Sirius, Rigel, Aldebaran, Capella, Castor and Pollux, and Procyon marking the angles, and Betelgeuse glittering not far from the centre of the figure. Observe Regulus with the "Sickle" of Leo on the meridian. Look for the glimmer of the "Beehive" in Cancer, between Gemini and Leo, and for the pentangular head of Hydra beneath it. Still lower you will see the reddish gleam of the starry serpent's heart, Cor Hydræ, or Alphard, and then, running eastward, and dipping ever nearer the horizon, the long, winding line of his stars passing under the overset cup of Crater and the quadrilateral of Corvus, the "Crow," until they disappear, unended, in the southeast, for from mid-heaven to the horizon there is not space enough to display all of these beautiful coils, which take a kind of life as you watch them.

Away over in the east, close to the ecliptic, you will see Virgo with her diamond, Spica, flashing in her hand. You are now facing east; to your left, then, north of Spica, glows great Arcturus, with his attendants shaping the figure of Boötes. Of Arcturus, a star that among a million finds no rival, we shall speak more particularly elsewhere. Farther to the left, beyond Boötes, shines the exquisite "Northern Crown," Corona Borealis. That too will claim attention in a later chapter. The square of Hercules is just above the horizon below the Crown in the northeast, and to its left, as you face north, is seen the diamond-shaped head of Draco, the "Great Dragon" that Athena was fabled to have entangled with the axis of the world. His stars wind upward between the "Dippers"—the "Little Dipper," which has the Polestar at the end of its handle, and the "Great Dipper," which, brim downward, shines east of the meridian, almost as high as the zenith, if you are as far north as 40° or more. The handle of the "Great Dipper" is the tail of Ursa Major, who treads lumberingly about the pole, with his back downward, his head out-thrust west of the meridian, and his feet, marked by three striking pairs of stars, up in the middle of the sky. On the meridian south of Ursa Major stands the "Sickle" of Leo already mentioned. Away round in the northwest, beyond Capella, are Perseus and Cassiopeia, immersed in the Milky Way.

Having fixed the location and general appearance of all these constellations in the mind, you are prepared to study them, and their stars, in more detail. Let us begin in the east. For some occult reason the rising stars always seem more attractive than those that are near setting. In the east, then, the eye is at once drawn to the beautiful Spica, which the impassive, immemorial Virgo wears as her only ornament. It is a fascinating star with its pure white rays, dashed with swift gleams of exquisite color as the atmospheric waves roll over it. There is not another equal to it in the impression of purity which it gives. We may imagine that some dim sense of this immaculate quality in the light of Spica led to the naming of the constellation the "Virgin," thus called by nearly all peoples, each in its own language: Π , Kó , *Puella, Kauni, She-Sang-Neu, Pucella, Vièrge, Mæden, Jungfrau, Virgine*—all, ancient and modern, Greek, Roman, Indian, Chinese, Norman, French, Anglo-Saxon, German, Italian, and English worshipping together at this shrine of ideal purity. If the Assyrians made her the wife of Bel that was hardly a disparagement, for Bel was the sun. So, too, the identification of Virgo with the Greek Persephone, the Roman Ceres, and the Jewish Bethula, all goddesses concerned with the harvest and the fertility of the land, in no way detracted from her virginal character, nor did her association with Astræa, the goddess of justice.

Beside Spica, Virgo has no very bright stars, and it is hardly doubtful that the imaginary purity ascribed to the constellation was derived entirely from the unsullied whiteness of Spica. While gazing at that beautiful star all of these associations, coming from times so remote and peoples so distant, crowd into the mind, increasing the interest with which one regards it. The nations who named it the vernal star, before all others, have gone the way of terrestrial things, but the star remains, as pearly fair as when Aratus sang to it:

"Lo, the Virgin!...
Her favor be upon us!"

Then science comes to carry the thoughts to grander, if less romantic, heights. Spica, it tells us, is a sun which might well claim to be included in Newcomb's wonderful "XM" class—*i. e.*, stars excelling our sun at least *ten thousand times* in splendor, for, notwithstanding the brilliance with which it delights us, it is so remote that no certain estimate of its distance can be made, its parallax escaping measurement—what, then, must be the intolerable blaze with which it illumines its immediate neighborhood! But when Science begins her revelations no man can foretell the wonders that she will discover. The spectroscope avers that Spica is speeding hitherward at a pace of more than 32,000 miles per hour! Each night that star is almost 700,000 miles nearer than it was the night before, and yet it is not perceptibly brighter than it was in the days of Homer. Such are the star depths! Such is the measureless playground of the spinning suns! Then Science, inspired by its spectroscopic sibyl, whispers another startling word in our ears: That core of white fire glowing so softly in the vernal midnight has an invisible companion star, with which it circles in an orbit 6,000,000 miles in diameter, and every four days they complete a swing in their mighty waltz!

The star Epsilon ( ) in Virgo (see Chart VII, at the end of the book) is *Vindemiatrix*, the "Grape-gatherer," thus named from some imagined association with the vintage. *Mukdim-al-Kitaf*, "The Forerunner of the Vintage," the Arabs called it, taking their hint from the Greeks before them.

Admiral Smyth, in his *Cycle of CelestialObjects*, has these curious lines on this star:

"Would you the Star of Bacchus find on noble Virgo's wing,

A lengthy ray from Hydra's heart unto Arcturus bring;

Two-thirds along that fancied line direct th' inquiring eye,

And there the jewel will be seen, south of Cor Caroli."

The reader may be interested in trying the star-loving admiral's plan for finding *Vindemiatrix*.

Gamma ( ) is *Porrima*, a prophetic goddess of ancient Latium, consulted especially by the women. But for us this star is most interesting as being one of the first binaries discovered in the heavens. It is a charming object for a small telescope. The two components revolve round their common centre of gravity in a period of about one hundred and eighty years.

As the reader progresses in his studies he will find Virgo full of interesting objects, including the celebrated "Field of the Nebulæ," marked out by the stars Beta ( ), Gamma ( ), Delta ( ), Epsilon ( ), and Eta ( ); but to see the nebulæ, which are thickly scattered there, he must have a powerful telescope.

Southwest of Virgo, but near the southeastern horizon, the quadrilateral figure of the constellation Corvus, the "Crow," catches the eye. Its brightest star is of less than the second magnitude, yet by their apparent association the four stars immediately attract attention. One sees no special reason why the figures marked out by these stars should be likened to the form of a bird; but it was a raven to both the Greeks and the Romans, and similarly symbolical to other early peoples. The Arabs, however, at first called it the "Tent," a designation which at least had a real resemblance for its basis. But these stars possess a charm independent of any fancied likeness to terrestrial things. In looking at them we do not think of the billions of miles which actually separate them from each other, but only of their seeming companionship. If, on the other hand, we force ourselves to consider the immense distances between them the mind is overwhelmed with the reflection that here, plainly staked out before us, is a field of space of absolutely unthinkable magnitude with its angles as clearly marked as if a celestial surveyor had placed corner-stones there. Note that the star Alpha ( ), once the leader of the constellation in brightness as well as in alphabetical rank, is now so faint that you have to look for it where it shrinks, in half concealment, below one of its now brighter neighbors. These abasements are not very uncommon among the stars. Their glory, too, is mutable; they also have their ups and downs. The Arabic name for Alpha ( ) was *Al Chiba*, or *Al Hiba*, meaning the "Tent." Gamma ( ), now the brightest star of the constellation, was called *Gienah*, the "Wing," and Delta ( ), *Algorab*, or *Al Ghurab*, the Arabic name for "Raven," but Beta ( ), which is perhaps as bright as Gamma ( ), has no special designation.

From Corvus the eye wanders naturally to its neighbor on the west, Crater, the "Cup." Both of these constellations rest on the back of the long serpentine Hydra. Crater is far less conspicuous than Corvus; but its resemblance to a cup is rather striking, although the imaginary vessel lies tipped up on its side with the open part toward the east. Among the many ascriptions of this starry cup in ancient mythology to various gods and goddesses, none is more interesting than that which made it the cup of Medea, thus including Crater among the numerous constellations which were associated in the imagination of the Greeks with their great romance of the Argonautic Expedition. Its brightest stars are only of the fourth and fifth magnitudes, and individually not worth much attention.

Hydra, which stretches its immense coils across about seven hours of right ascension, passing under Cancer, Leo, Crater, Corvus, Virgo, and a part of Libra, also carries the mind back through the golden mists of the morning of Greek mythology to the adventures of Jason and his crew of Argonauts, for it was once identified with the Aonian Dragon. It would be interesting to inquire how much of the perennial fascination of this ancient romance may be due to its traditional association with the stars. Look first at the head of Hydra, now well west of the meridian, below the glimmering "Beehive" in Cancer. It is marked by five stars of various magnitudes making an irregular pentagon. Then let the eye follow the line down southeastward until it encounters Cor Hydræ, or *Alphard*, the latter its Arabic name, meaning the "Solitary One." It is of the second magnitude and of a reddish color, and the space about it is vacant of conspicuous stars. There is an attraction about these solitary bright stars that is almost mystical, their very loneliness lending interest to the view, as when one watches some distant snow-clad peak gleaming in the rays of sunset after all the lower mountains have sunk into the blue shadows of coming night. Cor Hydræ is the Alpha ( ) of its constellation.

Above Hydra, northeast of Cor Hydræ, at the crossing of the ecliptic and the meridian, is the great star Regulus in Leo, the "Lion." It stands at the lower end of the handle of a very distinctly marked sickle-shaped figure, which includes the breast, head, and mane of the imaginary lion. Regulus is not only a beautiful star, but it possesses much practical importance as one of the principal "nautical stars," having been employed by sailors ever since the beginning of navigation to determine their place at sea. The sun almost runs over this star about the 20th of August, and every month the moon passes close beside it, and sometimes occults it. Thus it serves as a golden mile-stone in the sky. It has strangely affected the imagination of mankind in all ages. From the remotest times it has everywhere been known as the "royal star" *par excellence*. In Greek it was   , in Latin *Rex*, from which Copernicus constructed our name, Regulus. There are three other "royal stars," Aldebaran, Antares, and Fomalhaut, but Regulus has always been, in a certain way, their chief. For five thousand years it has been believed, traditionally, to control the affairs of heaven, and the astrologers have seized upon this idea by making it the natal star of kings, and those destined to kingly achievements and rule. In our age of science we may safely indulge these fancies; they can now do no harm, and they add immensely to the interest with which we regard the star that gave birth to them. When the "Royal Star" crosses high on the meridian in the vernal evenings, the imagination is thrown back over almost the whole course of the history of the Aryan race, and the rays of Regulus bring again the dreams of Babylon and Nineveh, of Greece and Rome, of India, and of the star-watching deserts of Arabia. Cyrus, in his conquering marches, may have looked to that star for help and inspiration, for it was the heavenly guardian of the Persian monarchs.

The spectroscope tells us that Regulus, like Spica, is approaching us, but less rapidly, drawing nearer about 475,000 miles per day. But its distance is 950,000,000,000,000 miles (parallax 0 .02), and it outshines the sun one thousand times.

The second star above Regulus, in the curve of the sickle's blade, is Gamma ( ), or *Algieba* (Arabic the "Forehead"), a beautiful double, probably binary, with a period of revolution which Doberck has estimated at about four hundred years. The larger star of the pair is golden-orange and the smaller bronze-green, a marvellous contrast, and an ordinary telescope shows well the spectacle, the distance between the components being 3 .78. And this wonderful pair is rushing toward the solar system at the rate of *two million miles per day.* Yet so great is its distance that we have no record that in a thousand years men have noticed a brightening of the headlight of this terrible locomotive of space! But probably the more refined methods of the present time, if applied for a similar period, would reveal an ominous expansion of that oncoming light. Gamma is interesting as marking, roughly, the spot in the sky which was the apparent centre of radiation for the November meteors, which were last seen in their splendor in 1866-67, their return in 1899-1900, for which the world had long been waiting, having been prevented by the disturbing attraction of Jupiter and Saturn, which shifted their orbit.

The "Sickle" in its entirety is an attractive asterism, and hanging so conspicuously in the sky on a spring evening it may be imaginatively regarded as a harbinger of the opening of the season when the thoughts of men are turning to preparations for future harvests. In the height of the harvest season the "Sickle" sets near sundown, then no longer standing upright, but lying along upon the horizon—a symbol of the wearied husbandman's approaching hours of rest:

"Nor shall a starry night his hopes betray."

Away off at the eastern end of the constellation, in the lion's tail, shines its second star in rank, Denebola (Arabic *Al Dhanab,* the "Tail"). It too is speeding hitherward, but only half as fast as Gamma. Like Aldebaran, the name Denebola has an indefinite charm, from its full round vowel sounds, and a certain nobility in the look of it as it lies on the printed page. As with many sonorous Indian names in American geography, these old star names lose something of their effect when they are translated. It is better to take them as they stand, transcending terrestrial analogy and definition, like the sublime objects that they designate.

Northeast of Denebola lies the small constellation of Coma Berenices, "Berenice's Hair," remarkable for the confused glitter of the small scattered stars of which it consists. It is a constellation with a romantic history which I shall not retell here. It forms an attraction for an opera-glass.

We now return to the region of sky above the head of Hydra, west of the meridian. There the attention is arrested by a glimmering spot, a kind of starry cobweb, which represents the "Beehive" cluster in Cancer. Its classical name is *Præsepe*, the "Manger." In *Astronomy with the Naked Eye* will be found a copy of Galileo's drawing of the stars of Præsepe as they appeared to him with his newly invented telescope. It is delightful to look at them on a clear night with a large opera-glass or a small telescope. They are an example of that clustering tendency so often seen among the stars, and which reaches its most wonderful manifestations in such assemblages as the famous globular clusters in Hercules and Centaurus, where countless thousands of small stars appear to be so crowded together that in the centre they run up into a perfect blaze. But in Præsepe there is no such apparent crowding, though the stars are so numerous that they resemble a swarm of bees. The probability is that none of the stars in this company is as large as our sun—although we cannot be perfectly sure because we do not know their distance—but they are, nevertheless, true stellar bodies, solar children, which seem playing together, overwatched by larger stars, waiting not far away. Plato, or his disciples, taking the suggestion from older dreamers, regarded Præsepe as a gateway of souls through which descended the spirits that were to animate the bodies of men during their earthly life. There are moods in which one can hardly consider our coldly scientific way of treating such celestial wonders as being essentially superior to the more spiritual ideas and suggestions of the visionaries of antiquity, before man became possessed with the notion that all science is summed up in measurement. Unquestionably we have more "facts," but have we more inspiration? Are we as near to the stars as were those who knew less about them? Have we yet got the key to unlock the universe? Do many of us comprehend the *dictum* of one of our own modern sages—"Hitch your wagon to a star"?

Cancer has no conspicuous stars, and it covers but a small space on the sky, yet as a constellation it is as old as any, and it has given us our "Tropic of Cancer," because in ancient times, before the Precession of the Equinoxes had drifted the zodiacal signs and constellations apart, the place of the Summer Solstice, where the sun is at its northern extreme of declination, was situated in Cancer, though now we find it in Gemini, close to the borders of Taurus.

Westward from Cancer we see the great group of mighty stars and constellations of which Orion is the chief and centre, but Sirius the brightest jewel. They are now declining rapidly toward the horizon, and will be better studied at another season. They include, besides Orion, Gemini, Auriga, Taurus, Canis Major, and Canis Minor, and will be found more favorably situated in the chart devoted to the sky at the Winter Solstice. For the present, then, we turn our eyes to the northern central part of the vernal heavens. There, almost overhead, shines the "Great Bear," Ursa Major, always recognizable by the remarkable figure of the "Great Dipper," or, as they prefer to call it in Old England—where brimming dippers of sparkling water lifted dripping from the "old oaken bucket" are not so familiar as in New England—the "Wain," or the "Plough." We have already remarked that at this season the Bear has his feet uppermost in the middle of the sky and his back downward toward the pole. The Dipper, too, is now upside down, drained of its last imaginary drop, though its stars may be the more brilliant for that. The figure of the bowl is situated on the flank of the Bear, and its handle represents his impossible tail. Six of its stars are of the second magnitude, and one, at the junction of the bowl and the handle, of the third. Their Greek letters, beginning at the northwestern corner of the bowl, are, *Alpha* ( ), *Beta* ( ), *Gamma* ( ), *Delta* ( ), *Epsilon* ( ), *Zeta* ( ), and *Eta* ( ), and their names, in the same order, *Dubhe, Merak, Phæd, Megrez, Alioth, Mizar,* and *Benetnasch.*

I once knew a country school-teacher who thought that he had acquired a pretty good knowledge of astronomy when he had learned these names by heart. He certainly knew more of uranography than most people. The names seem to be all of Arabic origin, and at the risk of destroying their charm I will give, from Allen's *Star Names,* their probable significations. *Dubhe* means simply "Bear"; *Merak* (sometimes *Mirak*), "Loin"; *Phæd* (sometimes *Phecda* or *Phad*), "Thigh"; *Megrez,* "Root of the Tail"; *Alioth,* meaning uncertain, probably something to do with the tail; *Mizar* (originally *Mirak*), "Girdle"; and *Benetnasch* (sometimes *Alcaid*), "Chief of the Mourners," from an Arabic phrase having that signification.

The star Megrez, now so much fainter than the others, was once as bright as any of them. It has faded within three hundred years.

Close by Mizar a fairly good eye has no difficulty in seeing a small star which is named Alcor (signification uncertain). The Arabs are said to have called these two stars the "Horse and his Rider," and to have regarded it as a test of good vision to be able to see them both. It is certainly not a severe test at present. Mizar itself is telescopically double, presenting a beautiful sight in a small telescope, the distance asunder being about 14 . The smaller star is like an emerald in hue, and the color is usually remarked at once by the beginner in telescopic observation. The larger star is one of those strange objects called "spectroscopic binaries"—two suns locked in the embrace of gravitation and spinning round a centre so near to each other that to anything less penetrating than the magic eye of the spectroscope they appear as a single body.

Merak and Dubhe are the celebrated "Pointers," so called because a line drawn from the former to the latter, and continued toward the pole, passes close to Polaris, the Pole-star, of which we shall presently speak. The distance between these stars is about five degrees, so that they serve as a rough measuring-stick for estimating distances in the sky. Immediately west of the meridian will be seen a curving row of stars which indicate the head of the Bear. Three of his feet, or claws, are represented by as many pairs of stars between the Great Dipper and the Sickle of Leo, one of the pairs being east of the meridian, one west of it, and one nearly upon it. Below the outer end of the handle of the Dipper, in the direction of Denebola, a fairly bright star, Cor Caroli, which English loyalty named for the heart of the unfortunate King Charles I., shines on the collar of one of the "Hunting Dogs," Canes Venatici, which Boötes is represented as holding in a leash as he chases Ursa Major round the pole. This, too, is a beautiful double, the contrasted colors of whose widely separated stars are finely shown by a small telescope.

Now let the eye run along the curve of the Dipper's handle, beginning at the bowl, and then, springing on, continue the same curve eastward; it will encounter, at a distance about equal to the whole length of the Dipper, a very great and brilliant star—Arcturus, brighter than Spica and Regulus, and usually, when not very far risen from the horizon, of a distinctly reddish hue. It is the chief star of Boötes, the "Driver," the "Vociferator," the "Herdsman," or the "Bear-watcher," as it has been variously rendered. We shall have more to say about Boötes in another chapter, but Arcturus is a star so splendid and famous that it cannot be passed in silence the first time the beginner catches sight of it. There is a standing dispute concerning the relative rank in brightness of Arcturus among the leading stars of the northern hemisphere. Its principal rivals are Vega in the Lyre, and Capella in Auriga. But all three differ in color, and that makes it more difficult to decide upon their relative brilliance, since different eyes vary in their sensitiveness to color. The Harvard Photometric *Durchmusterung* gives Vega the first and Arcturus the third rank among these three; but many eyes recognize rather a pre-eminence of Arcturus. My own impression has usually been that Arcturus looms larger than either Vega or Capella, but that Vega is the most penetratingly brilliant. It is very curious to notice the effect of the colors of these stars. The sharp blue ray in the light of Vega gives it a diamond-like quality which is lacking in Capella, whose light is white with just a suspicion of amber. Arcturus is a very pale topaz when high in the sky, and a ruddy yellow, sometimes flaming red, when near the horizon. It is a thrilling recollection of the writer's early boyhood that he felt an undefined fear of Arcturus when seen rising ominously red and flashing through the leafless boughs of an apple orchard in the late evenings of February. All the ancients feared Arcturus for its supposed influence in producing storms and bad harvests.

Arcturus is a sun of enormous magnitude, estimated all the way from one to six thousand times as great in luminosity as our sun. It is also travelling with great rapidity, its speed, according to some estimates, amounting to two or three hundred miles per second; but most of this is cross-motion with reference to us, its general direction being toward the south-southwest. If it is travelling three hundred miles per second, it would traverse the space between the sun and the nearest star, Alpha Centauri, in about three thousand years. We shall touch on Arcturus again when dealing with Boötes in the next chapter.

Disregarding for the present the exquisite circlet of Corona Borealis, the "Northern Crown," and the quadrilateral figure in Hercules, seen northeast of Arcturus, we turn to the great dragon, Draco, whose diamond-shaped head may be seen far over in the northeast above sparkling Vega, which is just on the horizon. As a reference to the charts of the circumpolar stars at the end of the book will show, Draco is a remarkably crooked constellation, its line of stars winding round between the "Little Dipper" in Ursa Minor, which has Polaris at the end of its handle, and the "Great Dipper" of Ursa Major. Its most interesting, though not now its brightest, star is Alpha, or *Thuban*, Arabic for "dragon." It lies between the end of the handle of the Great Dipper and the bowl of the small one. About forty-six hundred years ago Alpha Draconis was the Pole-star, and is believed to have shone down the long tube-like passage in the great pyramid of Cheops into the watching eyes of the priestly astronomers, assembled to view it in the mysterious chamber hollowed out of the solid rock deep under the foundations of the mighty pile. They thus had a telescope more than three hundred feet long as immovable as the solid earth, but, alas for their calculations, the star itself shifted its position, and their gigantic observing tube became useless until modern science inferred from its position the date of their building. How imposing to the imagination this association between a particular star and the mightiest structure made by human hands on the earth! Two centuries ago Thuban was more than twice as bright as it is now, and when the Egyptian priests sedulously observed it from their gloomy cavern, more than a thousand years before the magic-working days of Moses, it may have been brighter still.

Gamma ( ), or Eltanin (the "Dragon"), in the triangular head, is now the brightest star in the constellation, and it, too, has a history. Lockyer and others have identified it as the orientation star of Rameses' great temple at Karnak, and of the temples of Hathor and Mut at Dendera and Thebes. There is something magnificent in this thought of the ancient temple-builders—to square their work by the stars, and to construct long rows of sphinxes and majestic columns to conduct a ray from the sky to the eye of the god in his dark and hidden chamber, where no impious foot dared follow.

When you are tired of tracing the windings of the Dragon, turn to Ursa Minor and Polaris. The "Little Bear," it has been remarked, has an even more preposterous tail than his greater brother. Polaris is at the end of the tail, or the end of the handle of the Little Dipper, and the bowl of the latter is on the bear's flank.

If one knows nothing else of uranography, one should at least know Polaris, the "North Star." To recognize that star is to be able to orient yourself wherever you may be in the northern hemisphere. A whole volume could be written on its connection with human affairs. For at least two thousand years it has been the cynosure of sailors, and of wanderers by land as well. You cannot be lost if you have Polaris to guide you. The magnetic compass varies and misleads, the sun and the moon change their places, all the other stars circle through the heavens, but Polaris is always there, shining over the pole of the earth, the image of steadfastness. Only the slow Precession of the Equinoxes affects it. At the present time it is within one degree and a quarter of the true pole of the heavens, and it is drawing nearer that point, so that in two hundred years it will be less than half a degree from it— less than the apparent diameter of the moon. The little circle that it daily describes in the sky may be disregarded, for it is hardly noticeable except with instruments; but it is easy to fix the star's position with considerable accuracy by simple observation. Note that the Great Dipper and the "W"-shaped figure in Cassiopeia are on opposite sides of the pole. When one is above, the other is below; when one is on the east, the other is on the west. Draw an imaginary line from the star Mizar in the Great Dipper to the star Delta ( ) in Cassiopeia and it will pass almost directly through the pole. Polaris is on that line, a degree and a quarter from the pole in the direction of Delta Cassiopeiæ. If the observation is made when Delta is above the pole and Mizar below it, Polaris will be on the meridian, or north and south line, a degree and a quarter above the pole; when Delta is west of the pole and Mizar east of it, Polaris will be a degree and a quarter west of the meridian; when Delta is below the pole and Mizar above it, Polaris will be on the meridian a degree and a quarter below the pole; and, finally, when Delta is east of the pole and Mizar west of it, Polaris will be a degree and a quarter east of the meridian. The intermediate positions you can easily deduce for yourself.

But Polaris will not continue to be the unerring guide to the north that it now is. The Precession of the Equinoxes is carrying the pole progressively westward in right ascension, so that Polaris will eventually be left far behind. But the motion of the pole is in a circle about twenty-three and a half degrees in radius, and it requires about 25,800 years to complete a revolution round this circle. Consequently, at the end of that period, Polaris will have come back to reign again as the North Star for many centuries. In the interim other stars will have occupied its place. About 11,500 years from now the brilliant Vega, or Alpha Lyræ, will be the North Star, and in about 21,000 years Alpha Draconis (Thuban) will once more shine down the great northward-pointing passage in the pyramid of Cheops, if that pyramid shall still exist.

Polaris, unlike some of the others stars that we have been looking at, is running away into space instead of approaching us, at a speed which has been estimated at about 1,380,000 miles per day. Its present distance is not less than 200,000,000,000,000 miles. It has an invisible companion with which it circles in an orbit of a few million miles diameter in a period of about four days.

Polaris is also a celebrated visual double. With a telescope of two or three inches aperture you can see close by its flaming rays a minute blue star, a delicately beautiful sight. In the older days of telescopes, before they had attained the perfection which improvements in glass-making and lens-shaping have rendered possible, this little companion star of Polaris was a universal test of excellence. Its prestige was historical. The amateur owner of a telescope who could see that star clearly felt a joy that he could hardly express. The old makers of object-glasses, by rule of thumb, always tried them on the companion of the Polestar as a standard test for small apertures. The small star is of the ninth magnitude, and situated about 18 .6 from its primary.

The stars Beta ( ), or Kochab (the "Star"), and Gamma ( ), in Ursa Minor, are called the Wardens, or Guards, of the Pole. In low northern latitudes, where these stars sweep the horizon at their lower culmination, Shakespeare's description in *Othello* would be literally true during a great storm at sea:

"The wind-shak'd surge, with high and monstrous mane,

Seems to cast water on the burning Bear

And quench the guards of th' ever-fixed pole."

The constellations Cepheus, Cassiopeia, and Perseus, now low in the northwest and north, we leave for description to another chapter.

II THE EVENING SKY AT THE SUMMER SOLSTICE

At 10 o'clock P.M. on the 21st of June, the longest day of the northern hemisphere, the aspect of the sky is that shown in Chart II, accompanying this chapter. The same chart will answer for 11 P.M. on the 5th or 6th of June; 9 P.M. on the 7th of July, and 8 P.M. on the 22d or 23d of July. In fact, for any of the hours mentioned the date may be shifted several days forward or backward without seriously affecting the comparison of the chart with the sky, and the same may be said of each of the other circular charts. The stars simply rise about four minutes earlier each evening, and four minutes of time correspond to one degree of space measured on the face of the sky. So the whole sky shifts about one degree westward every twenty-four hours.

For the observation of the heavens at the epoch of the Summer Solstice, observers who are situated at least as far south as 40° north latitude have an advantage over those whose place on the earth is much farther north, because in the more northern regions sunset occurs later, and in England and Northern Europe the day, at this time, may exceed sixteen hours in length, while twilight is perceptible throughout the night. This interferes with the brilliancy of the stars.

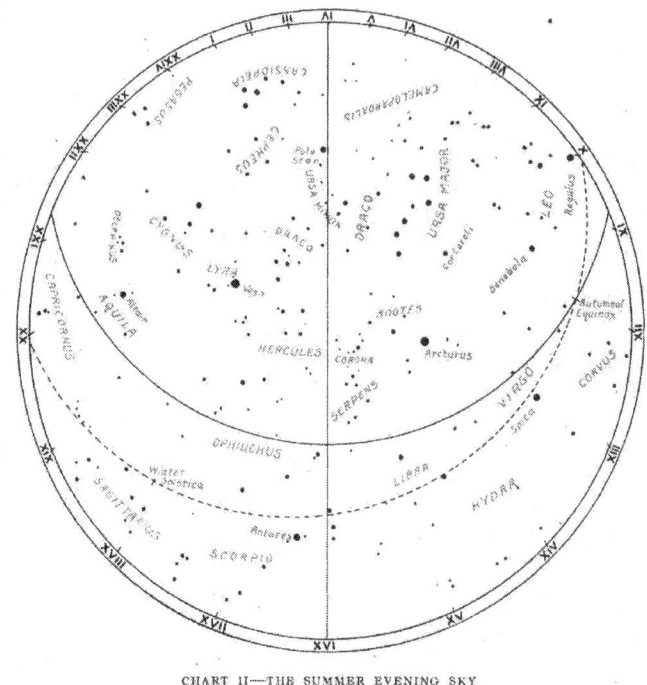

CHART II—THE SUMMER EVENING SKY

CHART II—THE SUMMER EVENING SKY

At no other season do the heavenly bodies seem so intimately associated with the earth as in summer. All nature is now attuned, and the stars glow softly in the tepid atmosphere, stirred by faint breezes, like veritable flowers of the sky. The firmament becomes a vast garden lit with beautiful lamps, which seem to have been placed there to dimly illuminate nocturnal wanderers in the transparent gloom beneath. Their beauty is as refreshing as the cooling breath of night itself. A mystic influence steals from them over the earth.

"If a man would be alone," says Emerson, "let him look at the stars."

Yet he cannot be alone with them; they are too friendly; they speak too plainly a universal language, which, though he cannot translate it, he *feels* in every fibre. There is nothing more absolutely common to all men than the influence of the stars. No one ever gazed up at them without feeling a change come over his spirit. Truly, "they separate between him and what he touches." They free him from the bondage of time and space. There is no trouble that they cannot assuage. And there is no time like the summer for becoming intimate with them. One who has been touched by the magic of their love could lie all the night long on a bed of pine-needles and fill his soul with their beauty. The march of red Antares and his glittering retinue across the meridian while the earth sleeps in solstitial calm—who can describe that pageant?

Antares is *the* summer star, and with it and the Scorpion we will begin. Not so bright as Arcturus or Vega, which are now high aloft, it has a charm peculiar to itself, arising partly from its fervid color, partly from its surroundings, and partly from its position, not too high above the southern horizon, which renders observation of the star comparatively easy. The color is so distinctive that one might think that he could recognize Antares chromatically if it were suddenly transported to some other region of the sky and placed amid a strange environment. Sometimes a flash of its fiery rays, striking sidewise into the eye as one is looking elsewhere, startles the observer like a red meteor. It is well named Antares—"Anti-Mars." With the telescope the wonder of color is increased, for close by the great star the glass reveals a smaller one of a *vivid green*, an all but incredible combination of complementarily tinted suns. And these suns are undoubtedly actually linked together into a system, so that, if there are planets revolving around both or either, the inhabitants of those planets may behold the spectacle of two suns, one crimson and the other emerald! The large star is of the first magnitude, and the small one of the seventh; angular distance 3 .7.

The companion of Antares is historically associated with the most interesting of American astronomers, a man whose life was a romance, Gen. O. M. Mitchel. When his long-cherished design of setting up a great telescope in America was at last fulfilled, at Cincinnati, in 1845, one of his first achievements was the discovery, to the surprise of the astronomers of Europe, of the green star hiding in the rays of Antares. At times it has been seen emerging from behind the moon, after an occultation, ahead of its red comrade.

With a parallax of 0 .02, Newcomb estimated the luminosity of Antares at nine hundred times that of our sun, and yet the spectroscope indicates that it is a dying sun, fast approaching extinction. In its younger days it may have been an orb of prodigious splendor.

The constellation Scorpio, of which Antares is the leader, is one of the best marked in the sky. The two small stars Sigma ( ) and Tau ( ), standing like attendants on either side of Antares, lend a singular aspect to the central part of the constellation. Antares is usually represented as the heart of the imaginary scorpion. Below Tau a curving row of stars dips to the southern horizon, and then rises, farther eastward, terminating with a conspicuous pair in the uplifted sting. West of Antares a nearly vertical row represents the head. Of the stars in this row, Beta ( ) is interesting as a fine and easily seen double, the distance being about 13 . A higher magnifying power shows that the larger star has another faint companion, distant only 0 .7. Nu ( ) is also telescopically interesting, for it consists of two pairs of stars. Observe in Chart VII the strange way in which the outlines of the constellation have been swung into loops in order to include certain stars in Scorpio, recalling the crooked boundary between Switzerland and Italy, by which each reserves particular peaks of the Alps for itself.

East of Scorpio, where the Milky Way, falling in flakes and sheets of silvery splendor upon the southeastern horizon, spreads abroad like an overflowing river, lies Sagittarius, the "Archer," often represented in the old pictorial charts as a centaur. The stars Lambda ( ), Delta ( ), and Epsilon ( ) form the bent bow. But modern eyes recognize more easily a dipper, formed by the stars Zeta ( ), Tau ( ), Sigma ( ), Phi ( ), Lambda ( ), and Mu ( ). But the star-clusters in Sagittarius are more interesting than the separate stars. A little southwest of Mu is the famous cluster 8 M., of which Barnard has made a photograph that is amazing beyond all description. Other clusters are all about in this part of the sky. A good opera-glass or field-glass is almost indispensable for one who would enjoy the glory of this wondrous region. Its riches are almost oppressive in their lavish abundance. Here one can have handfuls of stars for the picking up, like sands of gold from the bed of Pactolus. As the glittering incrustations that cover the roofs and walls of the Mammoth Cave are often compared to the starry heavens, so, reversing the image, Sagittarius is like a stupendous cavern of space all ablaze and aglitter with millions of sparkling gems.

Above Scorpio and Sagittarius are the intertwined constellations of Ophiuchus and Serpens. He who may wish to disentangle them is referred to *Astronomy with the Naked Eye*. But the outlines can be traced in Chart VII. The head of Serpens, like those of Hydra and Draco, is plainly marked by a striking group of stars, in this case resembling the figure called a "quincunx." From this point just under the "Northern Crown," the serpent's stars wind downward in beautiful pairs and groups, crossing the meridian above Scorpio, and rising again in the eastern part of the sky, above the little constellation of Sobieski's Shield, until they meet the borders of Aquila. Ophiuchus, with his head high up toward Hercules, where it is marked by the brightest star in that part of the sky, Alpha Ophiuchi, or Ras Alhague, the "Head of the Serpent Charmer," stands with legs braced wide apart, grasping the serpent at the points marked by the stars Delta ( ) and Epsilon ( ), and Tau ( ) and Nu ( ). It is Esculapius with his Serpent, said the Greeks; it is St. Paul and the Viper of Melita, or Moses and the Brazen Serpent, we don't know which, said the churchmen. I am not aware that in England they have ever been tempted to call it St. George and the Dragon. Politics and national pride have not meddled much with the stars, although there was once an attempt to fix the name of Napoleon upon Orion. Ras Alhague is described by R. H. Allen as sapphire in hue, while Alpha Serpentis is yellowish. The star Lambda ( ) in Ophiuchus, also called Marfik, the "Elbow," is a beautiful binary, period 235 years, distance apart 1 .2. The smaller star is smalt blue, a splendid telescopic object.

But, as in the case of Sagittarius, the greater wonders here are in the form of star-clusters, and particularly nebulæ. Just above Antares, in one of the feet of Ophiuchus, is a small star, Rho (to find which the reader must consult a large star atlas, like Klein's), around which Barnard has discovered, by photography, a truly marvellous nebula, a nebula which appears *to obscure the stars* like a cloud of cosmic dust. Great black lanes extend from and around it, and even the luminous parts of the nebula seem to absorb the light of the stars behind, diminishing their brightness a whole magnitude or more where they are veiled by it. This entire region of sky is most strange to the photographic eye. An outlier of the nebula just mentioned surrounds the star Nu ( ) in Scorpio, and its veiling effect upon the stars is even more evident. There is a similar appearance around the star Theta ( ) Ophiuchi, not far away. The sense of some appalling mystery in this part of the firmament is heightened by what Barnard says of a thing which has reappeared again and again on his photographs during the past fifteen years, at a point which he describes as lying very closely to R. A. xviii hours, 25 m., 31 s.; Decl. S. 26°, 9 (near the star Lambda ( ) in Sagittarius).

"It is a small, black hole in the sky. It is round and sharply defined. Its measured diameter on the negative is 2 .6. On account of its sharpness and smallness and its isolation, this is perhaps the most remarkable of all the black holes with which I am acquainted. It lies in an ordinary part of the Milky Way, and is not due to the presence or absence of stars, but seems really to be a marking on the sky itself" (*Astrophysical Journal*, January, 1910).

These things really transcend explanation (see *Curiosities of theSky*).

Above Ophiuchus and his Serpent, almost exactly overhead in the latitude of 40° N., we see the quadrilateral figure marked out by four of the principal stars of the constellation Hercules. The head of Draco, described in Chapter I, is beyond it toward the north-northeast. Hercules stands feet upward in the sky, his head, indicated by the star Alpha, or Ras Algethi, the "Kneeler's Head," being situated a few degrees west-northwest of Ras Alhague. Thus the two giants have their heads together. But while the occupation of Ophiuchus is plain, nobody, not even in ancient times, when the constellation received its name, has ever been able to say what Hercules is laboring at. When he was on the earth everybody followed his deeds and understood, if they could not emulate, them. He was as comprehensible as a modern pugilist. Now, however, that he has been translated to the stars, his labors are of a more mysterious nature, and, judging from his attitude, he finds them harder than any he undertook for the benefit of mankind here below. One is tempted to think that the powers he offended, when he boldly entered the land of shades and snatched the wife of his friend, King Admetus, from the hand of Death himself, are now taking an ample vengeance.

Ras Algethi is a very beautiful double star, one red, the other green or blue, and both, strangely enough, are variable in brightness. Their distance apart is 4 .7. Their spectrum indicates that they are advanced toward extinction many stages beyond our sun.

The star Zeta ( ), one of those in the quadrilateral, is a closer double, distance about 1 , and is binary, the period of revolution being about thirty-five years.

And now for a great marvel. Let the eye range slowly from Eta ( ) directly toward Zeta ( ). When one-third of the distance between the two stars has been passed, a faint, glimmering speck will be perceived. Perhaps you will need an opera-glass to make sure that you see it. This is the "Great Cluster in Hercules." You must go to the southern hemisphere to find its match anywhere in the sky. It is a ball of suns! Now you need a telescope. You *must have one.* You must either buy or borrow it, or you must pay a visit to an observatory, for this is a thing that no intelligent human being in these days can afford not to see. Can it be possible that any man can know that fifteen thousand suns are to be seen, burning in a compact globular cluster, and not long to regard them with his own eyes? Of what use is description in such a case? The language has not yet been invented to depict such things. Human speech comes down to us from the times when men did not need the tongue of the gods to tell what they saw. When Galileo invented the telescope, and Herschel multiplied its powers a thousandfold, they should have found a language fitted to describe their discoveries. But if you cannot get a look at the Hercules cluster through a powerful telescope, photography comes to your aid. Look at one of the wonderful Lick or Yerkes photographs of it, and pause long on what you see. Note the crowding of those suns toward the centre, note the glittering spiral lines formed by those which seem streaming and hurrying from all sides to join the marvellous congregation—and then turn again to that faint speck in the sky, which is all that the naked eye reveals of the wonder, and reflect upon the meaning of space and the universe.

We now turn farther east, still keeping the eyes directed high in the sky, and just at the edge of the Milky Way, with two minute stars making a little triangle with it, we see Vega or Alpha Lyræ, the astonishing brilliant that flashes on the strings of the heavenly Lyre. At the Vernal Equinox it was just rising far over in the northeast; now it is the unquestioned queen of that quarter of the sky. I like to think of Emerson when looking at that star. There is a sentence of his which reflects it like a mirror. When he strove to rouse the "sluggard intellect of this continent," to "look from under its iron lids," he could find no stronger image than that of poetry reviving here and leading in a new age, "as the star in the constellation Harp, which now flames in our zenith, astronomers announce, shall one day be the Pole-star for a thousand years."

Of the effect of the Precession of the Equinoxes, to which Emerson refers, we have already spoken. But it is a long time in the future that Vega will begin, or rather resume, its reign as the Star of the North. And, curiously enough, when that time comes the northern hemisphere will have its Summer Solstice when the sun is just opposite to the place which it now occupies at that season, and when Antares will be no more a summer star, but will flash its ruddy light upon the snows of a winter longer and colder than the winters that we know, while Orion will blaze above the summer landscapes. This immense revolution, some have thought, may be the measure of the "Great

Year" of Plato, and if the chronology adopted for dating the early remains of civilization recently uncovered in Crete is correct, we have evidence that mankind has persisted through one of these vast periods, and that nations flourished round the Mediterranean when Vega was formerly the Pole-star.

The beauty of Vega, which has been admired and commented on from the earliest times, is much enhanced when it is viewed with a telescope. Then the blueness of its light becomes evident, and one is the more astonished at the unquestionable fact that it outshines the sun a hundred times. A *sapphire* sun, a hundred times more brilliant than ours! The proper motion of the solar system, which carries us through space about twelve miles per second, is bearing us almost directly toward Vega, so that as future ages unroll the star should become brighter and brighter with decrease of distance, until eventually it may outshine every other orb in the firmament, and put Sirius himself to shame by its overpowering splendor.

The little star Epsilon ( ), the northernmost one of the pair near Vega, is a celebrated quadruple, easily seen as such with a telescope of moderate power.

A little less than half way from Beta ( ) to Gamma ( ) the telescope discovers the wonderful "Ring Nebula," a delicate circle of nebulous light with a star in the centre. This star is more conspicuous in photographs than in telescopic views. This object has been regarded as a visual proof of the correctness of Laplace's theory of the origin of the solar system from nebulous rings surrounding a central sun, but the Lick photographs show that the ring in this case is of a strangely complex constitution. Beta is both a binary and a variable star.

Buried in the Milky Way, east of Lyra, lies the great "Northern Cross" in the constellation Cygnus. It is more perfect than the famous "Southern Cross," and much larger. The star Alpha ( ), at the head of the main beam of the cross, is also called *Denib*, the "Tail," as it is situated in the tail of the "Swan," Cygnus. Its parallax is undetermined, and Newcomb placed it in his "XM" class, described under Spica in Chapter I. The Milky Way is exceedingly beautiful in Cygnus. Note particularly the broad gaps and rifts in it. Around and above the head of the cross there are dark spaces, which are specially impressive when the eyes are partly averted from them. Downward from Cygnus the stream of the galaxy is seen to be partially split longitudinally. It resembles a broad river meandering, in the droughts of the "dog days," over flats and shallows, and interrupted with long sand-bars. How can stars have been thrown together into such forms? What whirls and eddies of the ether can have made these *pools of shining suns?*

The star in the foot of the cross, Beta ( ), or *Albireo*—a beautiful name without signification, since Allen shows that it originated in a blunder (see his *Star Names and TheirMeanings*)—is one of the most attractive objects in the heavens for those who are fortunate enough to possess a telescope. The smallest glass easily shows it to be double, and the combination is unrivalled for beauty, the larger star being a pale topaz and the smaller a deep sapphire.

Their magnitudes are three and seven, and their distance apart about 34 . I have separated them with a field-glass.

Cygnus contains one of the nearest stars in the sky, a twinkler not too easily seen with the naked eye—a striking proof of the fact that the mere faintness of a star is in itself no indication of excessive distance. This is known as 61 Cygni, and will be found on Chart X. It is a double, possible binary, easily separated with a small telescope, the distance being about 21 . The distance of 61 Cygni is about 40,000,000,000,000 miles. It was long known as the second nearest star in the sky, the nearest being Alpha Centauri in the southern hemisphere; but at least one nearer one has more lately been discovered, and it, too, is a very small star. The combined luminosity of the two stars in 61 Cygni is only one-tenth that of the sun. Amid so many giants it is reassuring to find a sun smaller than ours; it restores our self-esteem to find that our solar hamlet is not the very least in the empire of space.

Southeast of Cygnus, near the eastern shore of the starry river, is Aquila, the "Eagle." Its chief star, Altair, "Eagle," recalls Antares, not by its color, for it is not red but white, but by the singular arrangement of two small stars standing one on either side of it. Here, too, the Milky Way is very splendid, attaining astonishing brightness lower down, in Scutum Sobieskii, "Sobieski's Shield." The naming of this constellation was a posthumous reward to the heroic king, John Sobieski, for saving Europe by the defeat of the Turks under the walls of Vienna, after their victorious advance from Constantinople, emphasized in the public mind by the appearance of Halley's Comet, had seemed to threaten a Moslem conquest. Twice Halley's Comet had alarmed Europe in connection with the Turks, first in 1456, after they had taken Constantinople, and again in 1682 when they swept upon Vienna, so that it was a natural thought to associate Sobieski's victory with some "sign in the sky," and a more appropriate one could hardly have been found than the "shield," bossed with star-clusters, which Hevelius selected for the purpose. The southern part of the constellation Aquila is sometimes called Antinous. For the beautiful Oriental legend of the Spinning Damsel and the Magpie Bridge connected with Aquila and Lyra, see *Astronomy with the Naked Eye*. Newcomb gives Altair ten times the luminosity of the sun.

The constellations Delphinus and Anser et Vulpecula will be dealt with in the next chapter. In the mean time let us turn to the western half of the sky.

Just west of the meridian, near the zenith, gleams the glorious Northern Crown, Corona Borealis. The head of Serpens is right underneath it. It is, perhaps, the most charming of all asterisms. It could hardly be called anything else than a crown or a wreath. The perfection of the figure is surprising. If its stars were larger it would be the cynosure of the sky, but small as they are they produce an effect of ensemble that could not have been exceeded if human hands had arranged them there. The superior brightness of one of them, Alpha "Gemma," or "The Pearl," adds greatly to the effectiveness of the combination. It is the work of a master jeweller! Yet, as I have elsewhere shown, this curious assemblage of stars is but a passing phenomenon, for they are travelling in various directions, with various speeds, and in the course of time the Northern Crown will dissolve like a figure in the clouds. In Greek mythology it was generally called the Crown of Ariadne. Just under the star Epsilon ( ) is a wonderful variable, which in 1866 suddenly blazed up to the second magnitude, and was for a time regarded as a new star. Nothing is known of its periods of change. It is not now visible to the naked eye.

West of Corona the most conspicuous object is Arcturus in Boötes. This entire constellation is now well placed for observation. But first a few words about Arcturus, a star of which one can never tire, so steeped is it in the poetry and history of the most interesting nations of the past. Like Alpha Centauri, Arcturus was used as a "temple star" in both Egypt and Greece, and it was of much importance as a prognosticator of the seasons. When a conspicuous star was seen rising just ahead of the sun, it was said to rise heliacally, and it served as a sentinel to announce the oncoming day. To the priests this was important, because it warned them of the moment when it was necessary to begin their preparations for the sunrise ceremonies in the temples. To the husbandman such a herald seemed specially connected with the particular season in which it appeared. In this way Arcturus came to give its name to the ancient Greek autumn. In Sophocles' *Œdipus theKing* there is a passage which affords striking evidence of the popular knowledge of Arcturus in this connection. When the herdsman from Mount Cithæron is brought to prove that he had nurtured Œdipus as a child, one of his former comrades, to recall the old man's recollections, reminds him that they had kept their flocks together "three whole half-years from Spring to Arcturus" (meaning from Spring to Autumn, since Arcturus then rose heliacally at the beginning of September). Whatever might be the local names for Autumn, over all the Greek world it was popularly known as the "time of Arcturus."

Although the Revised Version has struck out Arcturus and substituted "the Bear" in that famous passage in which the Almighty answers Job "out of the whirlwind," yet for lovers of the Bible this will always be "Job's Star," always surrounded to the imagination with the momentous circumstances suggested by that tremendous and unanswerable demand:

"Canst *thou* call forth *Arcturus* and his sons?"

No scientific fact known about it—not its gigantic size, not its inexplicable flight through space—can be so imposing as the impressions conveyed in its choice by Jehovah to illustrate His illimitable power. One likes to think that the Hebrew poet really did mean to write "Arcturus," for there is something sublime in the idea of representing the Great Maker of All as calling one of His stars by name.

Arcturus is sometimes referred to under the name of *Arctophilax*, the "Bear-driver," a name properly belonging to the constellation Boötes. In modern astronomical history it will always be memorable for the passage over it of the celebrated Comet of 1858, Donati's Comet. At one time the star was almost involved in the head of the great comet, and yet it shone through the obstructing vapors with virtually undiminished lustre. It was a spectacle, said Professor Nichol, the like of which no one might see again though he should spend on earth fifty lives. At the beginning the comet was a little plume of fire, "shaped like a bird of paradise," but it soon brightened into a stupendous scimetar, brandished in the sunset, and when it swept over Arcturus the whole astronomical world was watching to see what would happen to the star.

Among the other stars of Boötes, Epsilon ( ) is specially worthy of notice, being a remarkable binary of finely contrasted colors, orange and sea-green. The distance is 2 .25, and the period of revolution long but undetermined. Struve called this star "Pulcherrima," on account of its exceeding beauty.

Although Arcturus by its splendor belittles the rest of the constellation, yet it requires no difficult exercise of the imagination to see a giant form there, towering behind the Bear, and urging on his dogs in the chase. The dogs are represented by Canes Venatici, of the beauty of whose chief star, Cor Caroli, I have spoken in the preceding chapter. In the upper part of Canes Venatici, about 3° southwest of Benetnasch, is the celebrated "Whirlpool Nebula" of Lord Rosse, which modern photographs show in a form so suggestive of tremendous disruptive forces that cosmogonists are at a loss to explain it.

We now drop down to Libra, the "Balance," which lies just west of Scorpio and east of Virgo. There is evidence that this constellation originally represented the outstretched claws of the Scorpion. Yet as an independent constellation it is very ancient. It has only two stars of any considerable magnitude, Alpha ( ) and Beta ( ). The former must have faded, for it is now the fainter. It lies almost on the ecliptic. These stars are interesting on account of their curious names, which themselves tend to prove that Libra once formed a part of Scorpio. Alpha is Zubenelgenubi, the "Southern Claw," and Beta Zubeneschemali, the "Northern Claw." These titles, as Allen shows, have been derived through the Arabic from the Greek names current in the time of Ptolemy. The first is yellowish-white, and the second pale green. Any good eye detects the difference of color at a glance, although the stars are about ten degrees apart. Zubenelgenubi is widely double, separable with an opera-glass.

Along the western horizon we recognize our old friends Virgo, Corvus, and Leo, while high in the northwest is Ursa Major, head downward, and directly in the north Ursa Minor, standing on the end of his tail, poised like an acrobat on Polaris. The head of Draco shows finely east of the meridian, and low down in the northeast is the "Laconian Key" of Cassiopeia. But that is for another evening.

III THE EVENING SKY AT THE AUTUMNAL EQUINOX

"When descends on the Atlantic
  The gigantic
  Storm-wind of the Equinox,
  Landward in his wrath he scourges
  The toiling surges,
  Laden with sea-weed from the rocks."

Longfellow's vivid lines reproduce the popular impression of the character of the season when the descending sun again touches the equator, giving the whole world once more days and nights of equal length, before he dips to the south and leaves the northern hemisphere to face the oncoming blasts of winter. There is no superstition more deeply planted than that of the "equinoctial storms." There *are* such storms, it is true, but they by no means always burst at the epoch of the Equinox. The readjustment of atmospheric conditions goes on gradually, and there is often, just at the equinoctial moment, a spell of serene weather that can hardly be matched at any other season of the year. The atmosphere, recovered from the excessive heats of summer, possesses a quality of softness and "misty fruitfulness" that tranquillizes the spirit and makes nature doubly charming. It is the late afternoon of the year, when life, refreshed by the siestas of summer, resumes its activity, and the heavens no less than the face of the earth greet the eye with a smile of divine beauty.

To every season its flowers—and to every season its stars. The gardens of the sky are not the same in autumn as in summer, either in their arrangement or in the peculiarities of their bloom. There is less parade of flaming beauty, but the richness of the *coup d'œil* is not inferior. And just as in our September parterres some of the summer beauties remain, though a little faded, to support with their charms their stately successors, so in the skies of autumn a few of the summer stars are yet seen, though somewhat robbed of their

33

pristine splendor as they sink toward the sunset. The garland of the Milky Way has now been flung all across the firmament, from northeast to southwest, and while Vega and Altair hang half-way down the curtain of the west, recalling the glories of the solstice, Capella appears rising in the northeast, and Cassiopeia, not less beautiful in the sky than when she awoke the jealousy of the sea-nymphs, is seen seated in her "shiny chair" east of the meridian in the north. Between Cassiopeia and Capella flashes Perseus, with his uplifted sword marked by a curve of stars embedded in the Milky Way, and above Perseus stands Andromeda, upright, with her feet toward her rescuer and her head touching the "Great Square of Pegasus," near the middle of the sky, east of the meridian. Cepheus, the King, is on the meridian above the pole. Cassiopeia, Cepheus, Andromeda, and Perseus constitute the "Royal Family" of the sky, more enduring than the proud dynasties that by turns have ruled terrestrial affairs.

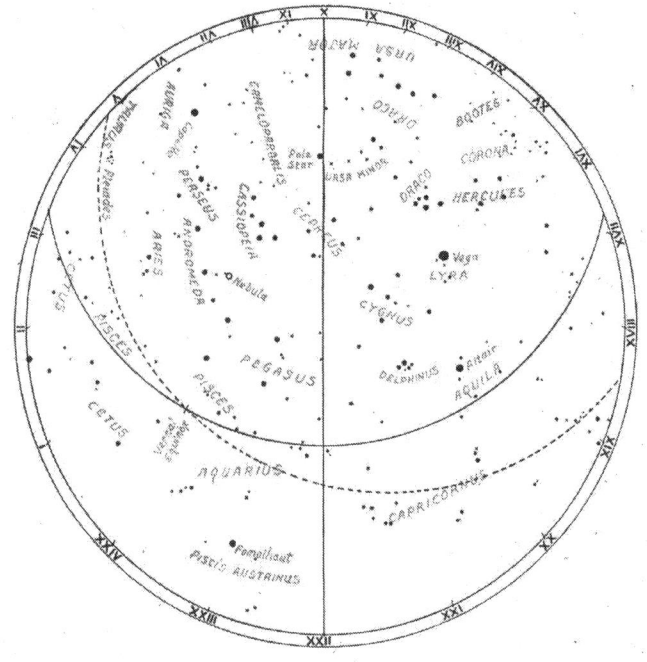

CHART III—THE AUTUMN EVENING SKY

CHART III—THE AUTUMN EVENING SKY

Low down in the south, east of the meridian, glows Fomalhaut, the "Fish's Mouth," the leading and the only bright star of Piscis Austrinus, the "Southern Fish." With this singular star we may begin our description of the beauties of the autumn sky. Fomalhaut well deserves the epithet singular, if for nothing else than on account of its loneliness. In this respect it is more remarkable than Cor Hydræ, which it resembles in its ruddy color. Fomalhaut is the characteristic star of autumn in our latitudes, for the same reasons that cause Antares to represent the summer. Like Antares, it startles the wandering eye and fixes the attention, although, unlike the great star of Scorpio, it has no brilliant *entourage* to emphasize its supremacy over the quarter of the sky where it shines. It is one of the sailors' stars. To me Fomalhaut is full of boyhood memories and impressions gained when I learned the stars in the country, among the hills that shut in the Schoharie before it pours out into the valley of the Mohawk. Fortunately, Thomas Dick's works and Burritt's *Geographyof the Heavens* had a place in our house, and neither *TheArabian Nights* nor *The Swiss Family Robinson* was able to dull my appetite for them. In the course of time I knew all the great stars by name, and found a wonderful pleasure in their acquaintance, although at times they daunted me with their imposing associations with Egypt, the Nile, Babylonia, and everything that is most ancient. I shall never forget Fomalhaut flashing along in the south, just skipping the hilltops on an autumn night. A great star is never so imposing nor so mysterious as when it thus appears to be watching the earth.

How immensely would the interest of many travellers' tales be heightened if only they had known the names of the stars whose appearance they have recorded. When you have the name of the star that was seen, the season and the hour of the night are fixed at once, and the whole scene is filled with new life. When an Alpine climber, waiting in his lonely camp high on the mountain-side for the coming of day, tells me, "I saw Sirius glancing at us over a lofty peak far in the east," I know immediately the approximate time of night and the aspect of the heavens, and the narrative gains in vividness; but if he says merely that he saw "a star" his stroke of description misses. And, then, the names of many of the stars, by their oddity and beauty, enrich the page and awake the imagination. They are, in themselves, an incantation.

The lover of the stars is grateful for any reference to them by a great writer, and yet he is often disappointed by the inadequacy of descriptions that might easily have been made memorable if only their authors had known the starry heavens a little better. How disappointing, for instance, is this passage in R. H. Dana's *Two Years before the Mast*:

"Wednesday, November 5th—The weather was fine during the previous night, and we had a clear view of the Magellan Clouds and of the Southern Cross. The Magellan Clouds consist of three small nebulæ in the southern part of the heavens—two bright, like the Milky Way, and one dark. They are first seen just above the horizon after crossing the southern tropic. When off Cape Horn they are nearly overhead. The Cross is composed of four stars in that form, and it is said to be the brightest constellation in the heavens."

That is all, and the reader's dissatisfaction is not confined to the evidence of the writer's lack of familiarity with the stars, but becomes yet keener when he reflects upon the brilliant picture which Mr. Dana's powers of description should have enabled him to make of those strange sights of the southern sky, which, in his day, were so rarely seen by northern eyes.

On the equator above Fomalhaut, and close to the meridian, appears a curious group of stars in the form of a letter Y. They mark the hand and urn of Aquarius, the "Waterman." A few degrees westward from this figure shines the Alpha ( ) of the constellation, bearing the strange name Sadalmelik, the "King's Luck," or "Lucky One." It is situated in the Waterman's right shoulder, while Beta ( ), some twelve degrees farther west, marks the left shoulder. Beta's distinctive name is Sadalsuud, the "Luckiest of the Lucky." Several other stars in this constellation have names implying good-fortune. The Arabs saw the Y-shaped figure, already referred to, as a tent, and the star Gamma ( ) in this group is called Sadachbiah, from an Arabic phrase which Professor Whitney translates "Felicity of Tents." Upon this R. H. Allen remarks that the star probably got its name from the fact that it rose with its companions in the morning twilight of spring, "when, after the winter's want and suffering, the nomads' tents were raised on the freshening pastures, and the pleasant weather set in." The star Zeta ( ), in this same figure, is a long-period binary, probably 750 years, and a beautiful telescopic object, the distance being a little more than 3 , while the two stars are nearly equal, and very white, although one of them seems whiter than the other.

It will be observed that the outline of the constellation Aquarius is very curious, somewhat resembling that of the State of Louisiana tipped on its side. The broader part of it runs down toward Fomalhaut, and the northern part extends westward, like an L added to a house, between Equuleus and Capricornus. The latter, the constellation of the "Goat," is relatively small and compact. Its two most interesting stars are Alpha ( ), or Algedi, the "Goat," and Beta ( ), or Dabih (signification uncertain), both in one of the horns of the imaginary animal. Each of these stars is a wide double. The distance between the Alphas is 373 , and that between the Betas 205 , the latter being more than a tenth of the apparent diameter of the moon. A good eye sees at once that Alpha is double; but the two stars in Beta cannot be seen without a glass, because one of them is below the sixth magnitude, the *minimum visible* for the naked eye. Each of the stars in Beta is a telescopic double. The Goat heads westward, and the stars Delta ( ) and Gamma ( ) are in his tail. This constellation has given us our Tropic of Capricorn, because the place of the winter solstice was once within its boundaries, although now we find it far west, in Sagittarius.

Above the head of Capricornus we recognize our old acquaintance Altair, in the Eagle, and east of this the singular little constellation of Delphinus, the "Dolphin," often called "Job's Coffin," a name for which I have never been able to find any explanation. Like all small constellations whose stars are comparatively close together, it immediately attracts the eye. None of its stars exceeds the fourth magnitude; but three of them, Alpha, Beta, and Gamma, are telescopic doubles, the last named being particularly beautiful on account of the contrast of colors, gold and green; distance 11 .

Directly north of Altair is the very small constellation of Sagitta, the "Arrow," interesting when viewed with an opera-glass for its row of little stars from which, as from a maypole lying horizontally, depend loops of still smaller stars looking like garlands. In ancient times this was sometimes called "Cupid's Arrow," but they did not venture to represent the little god himself. Above Sagitta are the small stars constituting the double constellation of Vulpecula et Anser, the "Little Fox and the Goose."

Simply pausing to recognize the presence of the Northern Cross, we turn to the eastern side of the meridian, where we find Pegasus, with his Great Square. This is one of the most conspicuous figures in the sky. The star at the northeastern corner of the square is Alpheratz, of which I have spoken in the Introduction, as belonging in common to Andromeda and Pegasus. When we come to Cassiopeia I shall point out a remarkable fact relating to Alpheratz and its twin, Gamma Pegasi, about 15 degrees directly south. Every lover of the "classics" of course feels a thrill of pleasure in seeing Pegasus in the sky, "in wild flight and free." One can spare many of the heroes for the sake of giving him room. Shakespeare's references to the constellations are much less frequent and definite than one could wish, but he has clearly mentioned one or two, and it may be that he had the starry eidolon of the Winged Horse in his eye when he wrote, in *Troilus andCressida*:

"But let the ruffian Boreas once enrage
The gentle Thetis, and anon behold
The strong-ribbed bark through liquid mountains cut,
Bounding between the two moist elements
Like Perseus' horse."

The constellation extends far westward from the Square, and in the imaginative sky pictures that illustrate old charts of the heavens the star Epsilon ( ) is in the nose of Pegasus, as he stretches out his neck to reach his foal, Equuleus. But the horse, with his feet toward the north, is shown upside down, unless you turn your back to the south when looking at him. The star Beta ( ) is attractive on account of its neighbors forming a striking triangle with it; but the space within the Square is relatively vacant. Alpha ( ) and Beta ( ) are respectively Markab, the "Saddle," and Scheat (signification uncertain).

South of the Square of Pegasus we see the western part of the constellation of Pisces, whose small stars run in streams toward the eastern horizon. Pisces furnishes one of the most remarkable examples of this phenomenon, in which the stars are seen arrayed in long, winding lines, like buttercups following a brook. Cetus is also seen rising south of Pisces; but we shall deal with these constellations later. Meanwhile we return to Alpheratz, at the northeast corner of the Square of Pegasus. The name is derived from an Arabic phrase meaning the "Horse's Navel"; but the star is now generally associated with Andromeda, and is, indeed, the Alpha of that constellation, and shines on the maiden's head. The star Delta ( ), in Andromeda, marks her breast, and her extended arms and chained hands are shown by rows and groups of small stars on the north and south. Beta ( ), or Mirach, is in her girdle, and the two small stars northwest of it lead the eye to one of the most wonderful objects in the sky—the Great Andromeda Nebula. You may detect it as a misty speck with the naked eye; an opera-glass will show you plainly that it is a little luminous cloud. In Chart X its position is indicated by a little circle near the star Nu ( ). In a telescope it appears of a spindle shape, with a bright axis, but the best views of it are afforded by photography. On the photographic plate, exposed continuously for hours to its rays, it gradually builds up its marvellous form— the great central condensation, with the encircling spirals, emerging in all their strange splendor. It resembles a whirlwind of snow, and the appearance of swift motion and terrific force is startling. Its spectrum, instead of being that characteristic of gases, indicates that it consists principally of matter in a star-like state of condensation, and some have imagined that it is an outside universe, composed of stars too distant to be separately distinguished, and arrayed in mighty spirals, which recall the form of the Milky Way. The latest investigations show evidence, however, that it is partly nebular in constitution. These things once known, the contemplative eye is drawn to that misty speck as to a magnet.

The star Gamma ( ), or Almaack, the "Badger," is in Andromeda's foot. It is a wonderful triple star, whose largest member is orange in color, the second emerald-green, and the third blue. The two larger stars are easily seen with an ordinary telescope, the distance between them being about 10 , but the third is difficult, the distance from the second being, in 1908, only 0 .45. The last two form a binary, with a period of about fifty-four years. When they are nearest to each other no telescope can separate them. The colors of the two largest stars are very striking, and yet some eyes seem incapable of appreciating them. This is also true of many separate stars in the sky which possess distinctive tints. It is a fine test of the chromatic capacity of the eye to be able to enjoy the differences among the hues of the stars. Color-blindness is far more common than is usually suspected, and is apt to manifest itself in this way when not otherwise noticed. From theoretical considerations Holmgren has shown that three varieties of color-blindness may exist: first, where the sense is defective for only one color, either red, green, or violet; second, where two colors, either red and green or red and violet, are not perceived; and third, where the defect extends to three colors, including red, green, and violet. A person suffering from either of these forms of blindness would lose much of the peculiar beauty exhibited by certain stars and combinations of stars.

To the right of Almaack, as one faces north, is the little constellation of Triangulum, and beyond that, in the same direction, Aries, the "Ram," clearly marked by three stars, the two smaller of which are quite close together. The largest star, Alpha ( ), is called Hamal, the "Ram," or "Sheep"; and the next largest, Beta ( ), Sheratan, the "Sign," this name being due to the fact that in the days of Hipparchus Sheratan marked the place of the Vernal Equinox, and consequently the point of beginning of the year, of which it was the sign. Gamma ( ), the companion of Sheratan, sometimes called Mesarthim (signification uncertain), is a beautiful telescopic double whose components are 8 .8 apart. The smaller one has a curious tint which Webb and others have described as "gray."

Aries was originally the leader of the zodiac, but the Precession of the Equinoxes has now thrown it into second place, and brought Pisces to the front, the twelve signs of the zodiac being like a fixed circular framework through which the constellations drift toward the east. The *sign* Aries remains the first of the zodiac, but is occupied by the constellation Pisces. Is there in any language a word more mysteriously impressive than "zodiac"? Astrological superstition, perhaps, partly accounts for this. The word comes from the Greek for "animal," because nearly all the constellations of the zodiacal circle are representations of animals. It surrounds the sky with a great menagerie of starry phantasms, through the midst of which the sun pursues his annual round. When he enters the sign of Aries spring commences; when he enters Cancer summer reigns; when he reaches Libra it is the beginning of autumn, and when he is in Capricorn winter is at hand. We have nothing quite equal to the old Greek story of Phaeton begging from his father, Phœbus Apollo, the privilege of driving the Chariot of the Sun, and losing his way through terror of the threatening forms amid which lay his course—the "Scorpion," with his fiery sting uplifted to strike; the huge "Crab," sprawling across the way; the fierce "Ram," with lowered head; the great "Bull," charging headlong upon him; the terrible "Lion," with bristling mane; the "Archer," with bow bent and arrow aimed; the "Goat," with crooked, threatening horns; the sturdy "Waterman," emptying his vast urn in a raging flood; the balance of "Libra" extended as if to weigh his fate—even the benign aspect of the "Twins" and the gentle look of the sedate "Virgin" could not restore his equanimity. It was the wildest of all wild rides, and Phaeton was the precursor of the modern chauffeur gone mad with the speed of his flight, and crazed by the pursuit of phantoms which rise remorselessly in his path. It was probably in Aries that the inventors of the story imagined the beginning of the adventure.

Below the feet of Andromeda, in the northeast, appears Perseus, her rescuer, hurrying to the combat with the oncoming Sea Monster, and carrying the blood-freezing head of Medusa in one hand and his diamond-hilted sword in the other. He wraps the glory of the Milky Way around him like a flying mantle, and brandished in the direction of Cassiopeia, the maiden's mother, and of King Cepheus, her father, is seen his magic blade, made splendid in the sky by one of the finest assemblages of small stars that can anywhere be seen. This beautiful star-swarm, visible to the naked eye as a glowing patch in the Milky Way, is indicated in Chart X by a double cluster of dots above the star Eta ( ). Seen with a powerful opera-glass, or better with a small telescope, it is an object that one can never cease to admire and wonder at. It is so bright that the unassisted eye sees it as soon as it is directed toward that part of the sky. It seems to throw a halo over the surrounding sky, as if at that point the galaxy had been tied into a gleaming knot. It is popularly called the "Sword Hand of Perseus." But how inadequate seems such terrestrial imagery when we reflect that here a vast chaotic nebula has been, through æons of evolution, transformed into a kingdom of starry beauty.

The star Alpha ( ) Persei, also known as Algenib (Arabic *AlJanib*, the "Side"), is the centre of a bending row following the curve of the Milky Way. The appearance of this curve of stars is very attractive to the eye. Algenib is a beautiful star, allied to our sun in spectroscopic character, and approaching us at the rate of about 560,000 miles per day.

But the greatest marvel of Perseus is the "Demon Star," Algol, in the head of Medusa, which is represented depending from the hero's right hand. Algol bears the Greek letter Beta ( ). It is the most wonderful of variables, and its variations can be watched without any instrumental assistance. For the greater part of the time it is of nearly the second magnitude; but once every two days, twenty hours, and forty-nine seconds it begins suddenly to lose light, and in about four hours or less it fades to nearly the fourth magnitude, being then no brighter than some of the faint stars around it. Almost immediately it begins to brighten again, and in the course of about three hours is seen shining with its pristine splendor. The cause of these singular variations is believed to be the existence of a dark star, or a mass of meteors, revolving round Algol at such close quarters that a distance of only 3,000,000 miles separates the centres of the two. Algol itself is demonstrably considerably larger than our sun, but of less density. The Arabic name for this star was *Al Ghul*, the "Demon," or "Fiend of the Woods," and our word ghoul comes from it. The imagination of a Poe could not have represented a more startling thing—a sun that winks like a gloating demon! One may easily cultivate an uncanny feeling while watching it. No one need be surprised that the astrologers make much of the malign influence of Algol. If one had faith in them, one might as well be born with the millstone of fate tied to his neck as to have Algol in his nativity.

Below Perseus, and not very high above the horizon, sparkles the brilliant Capella, but that is for the next chapter. We turn to Cassiopeia. Her "W," or "Laconian Key," is a familiar asterism to all who know anything at all of the starry heavens. The five stars forming this figure are also represented as marking the Chair in which the unfortunate though beautiful queen sits. There is a delightful reference to this "Chair" in Xavier de Maistre's *ExpéditionNocturne autour de ma Chambre.* When the hero discovers the slipper of his fair neighbor of the upper flat visible on the balcony above, he wishes "to compare the pleasure that a modest man may feel in contemplating a lady's slipper with that imparted by the contemplation of the stars." Accordingly, he chooses the first constellation that he can see. "It was, if

I mistake not, Cassiopeia's Chair which I saw over my head, and I looked by turns at the constellation and the slipper, the slipper and the constellation. I perceived then that these two sensations were of a totally different nature; the one was in my head, while the other seemed to me to have its seat in the region of the heart."

The names of three of the five stars forming the "Chair" are: Alpha ( ) Schedar (from *Al Sadr*, the "Breast"); Beta ( ) Caph (Arabic *Kaff*, "Hand"); and Delta ( ) Ruchbah or Rucbar, the "Knee." Caph and Ruchbar are of particular interest, the first because, together with Alpheratz and Gamma Pegasi (often called Algenib, although that name belongs to Alpha Persei), it lies almost exactly on the Equinoctial Colure, or First Meridian of the Heavens; and Ruchbah, because, as explained in Chapter I, it lies in a line with Polaris and the true pole, thus serving to indicate the position of Polaris with regard to the pole at any time. Caph, Alpheratz, and Gamma Pegasi are often called the "Three Guides," because, as just explained, they graphically show the line of the Equinoctial Colure, which is a great circle passing through the pole and cutting the equator at the Vernal and Autumnal Equinoxes. On the opposite side of the pole this line passes between the stars Gamma ( ) and Delta ( ) in Ursa Major.

The star Eta ( ) is an extremely beautiful binary, period about two hundred years, distance at present more than 6 . The combination of colors is especially remarkable, the larger component being orange, and the smaller purple. Piazzi Smyth saw the color of the smaller star as "Indian red," and others have variously called it "garnet," "violet," and, curiously enough, considering the general opinion to the contrary, "green." There is no doubt, whatever the exact hue may be, that this star wears a livery distinguishing it from any other in the sky. It is hardly an exaggeration to say that there is as great a variety of color tones among stars as among flowers. Although the great majority of stars approximate to white, there are, nevertheless, red stars, green stars, blue stars, lilac stars, yellow stars, orange stars, indigo stars, and violet stars, and stars of other tints and shades. All of those which are deeply colored are linked together in close pairs, but the colors they exhibit are not an effect of contrast. It is wonderful to think of *suns* of such hues, but *there they are*! And, after all, it would be no more difficult to account for the colors of stars than for those of flowers. But to live under a purple or an emerald sun might not be as agreeable as life in the rays of our white orb, whose light splits into rainbows, as light of a single primary color could not do. A flower-garden under a green sun would not be the marvel of prismatic hues that it is in our world.[1]

Cassiopeia is memorable for being the scene of one of the greatest astronomical occurrences on record. Near the star Kappa ( ), in 1572, appeared the most splendid new star that has ever been seen. It is known as "Tycho's Star," the Danish astronomer Tycho Brahe having been an assiduous student of the wonderful phenomenon during the sixteen months that it remained visible. There is a red variable star of less than the tenth magnitude quite close to the spot where Tycho recorded the appearance of his *nova*, and it has been thought that this may be the mysterious object itself. In 1901 a new star, almost equal in brilliance to Tycho's, suddenly burst out in Perseus, between Algol and Algenib, and these two so similar phenomena occurring in the same quarter of the heavens are usually linked together in the discussion of new stars. The reader who wishes more particulars about these stars may consult *Curiosities of the Sky*.

The background of the sky around Cassiopeia is a magnificent field for the opera-glass and the telescope. In sweeping over it one is reminded of Jean Paul Richter's *Dream of the Universe*:

"Thus we flew on through the starry wildernesses; one heaven after another unfurled its immeasurable banners before us and then rolled up behind us; galaxy behind galaxy towered up into solemn altitudes before which the spirit shuddered; and they stood in long array, through which the Infinite Beings might pass in progress. Sometimes the Form that lightened would outfly my weary thoughts, and then it would be seen far off before me like a coruscation among the stars, till suddenly I thought to myself the thought of 'There,' and then I was at its side. But as we were thus swallowed up by one abyss of stars after another, and the heavens above our eyes were not emptier, neither were the heavens below them fuller; and as suns without intermission fell into the solar ocean like waterspouts of a storm which fall into the ocean of waters, then at length the human heart within me was overburdened and weary, and yearned after some narrow cell or quiet oratory in this metropolitan cathedral of the universe. And I said to the Form at my side: 'O Spirit! has then this universe no end?' And the Form answered and said, 'Lo! it has no beginning!'"

Westward from Cassiopeia, directly over the pole, and lying athwart the meridian, is the constellation of Cepheus, the King, less conspicuous than that of his queen, Cassiopeia, but equally ancient. Its leading star, Alpha ( ), also called Alderamin, the "Right Arm," is a candidate for the great office of Pole-star, which it will occupy in about 5500 years. Beta ( ), the second in rank, is named Alfirk, the "Flock" or "Herd." If you are sweeping here with an opera-glass you will perceive, about half-way between Alpha ( ) and Zeta ( ), a small star which will at once arrest your attention by its color. It is the celebrated "Garnet Star" of Sir William Herschel, who was greatly impressed by its brilliant hue, declaring it to be the most deeply colored star that the naked eye can find in the sky. But its color is not so striking unless a glass be used.

Low down in the north-northwest we see the Great Dipper, above it the coiling form and diamond head of Draco, and then, still higher, the Northern Cross and Vega, bright as a jewel. Hercules and the Northern Crown are near setting in the northwest.

FOOTNOTES:

[1] The reader who is curious concerning such matters is advised to consult a paper by Dr. Louis Bell on "Star Colors," in the *Astrophysical Journal* (vol. xxi, No. 3, April, 1910). Dr. Bell's experiments with artificial stars seem to show that physiological effects play a great part in producing the pronounced colors of the small stars in many telescopic doubles. The paper is very interesting, especially in its description of a startling imitation of the singular cluster, Kappa ( ) Crucis, which Sir John Herschel described as resembling a gorgeous piece of colored jewelry. But, whatever part physiological optics may play in the phenomena of colored doubles, it is certain that many single stars, including some of great magnitude, possess distinctive tints. Compare, for instance, Castor and Pollux or Rigel and Betelgeuse. Aldebaran and Betelgeuse are both reddish, yet the color tones that they exhibit are clearly different.

IV THE EVENING SKY AT THE WINTER SOLSTICE

The magic of the starry heavens does not fail with the decline of the sun in winter, but, on the contrary, increases in power when the curtains of the night begin to close so early that by six o'clock the twilight is gone and the firmament has become a dome of jet ablaze with clusters of living gems. And when the snows arrive, mantling the hills with glistening ermine, the coruscating splendor of the sky seems to be redoubled. If I were to choose a time most suitable for interesting a novice in the beauties and wonders of uranography, I would select the winter, and I would lead my acolyte, on a clear, frosty night, when the landscape was glittering with crusted snow, upon some eminence where the curve of the horizon was broken only by the leafless tops of a few trees, through which the rising stars would flash like electric lamps. The accord between the stars and the seasons is never more evident than at such a time and in such a place, and the psychology of the stars is then most strongly felt. When the earth is locked fast in the bonds of winter the sparkling heavens seem most alive. I would have, if it were possible, a clump of dark pines or hemlocks near the place of observation, throwing their shadows on the snow, while Sirius in all its wild beauty blazed above them, and Aldebaran, Rigel, and Betelgeuse filled the vibrant air about them with jewelled lances of prismatic light. Then the sound of sleigh-bells in the resonant atmosphere would seem an aerial music shaken from the scintillant sky, and a lurking fox, stealing from his den in the edge of the shadows, would appear timorously conscious of the splendor over his head. The nocturnal animals know a day more glorious than ours, but it is never so glorious as when its multi-colored rays splinter upon crystalled hills at the winter solstice.

Now the greatest of the constellations reign in the sky. Orion is high up in the southeast, and around him are arrayed his brilliant attendants and companions—toward the west Taurus, with Aldebaran and the glittering Pleiades; above, Auriga and Gemini dipping their feet in the Milky Way; in the east, Canis Minor, with great, steady Procyon, and Canis Major proclaiming his precedence with flaming Sirius, the King of the Stars. We cannot do better than begin with this starry monarch and his constellation.

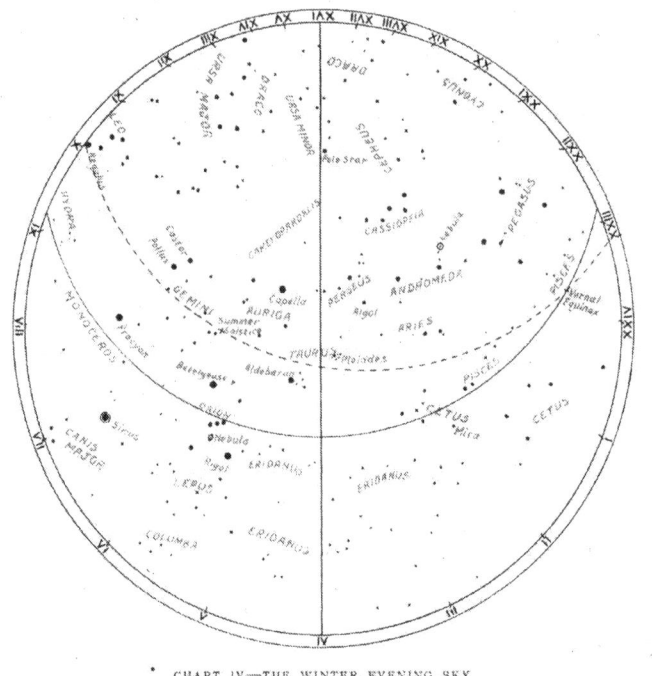

CHART IV—THE WINTER EVENING SKY

CHART IV—THE WINTER EVENING SKY

To me Sirius will always remain associated with the memory of Christmas sleigh-bells and the thrilling creak of runners on crisp, hard snow, for it was during a drive home from a "Christmas-tree" in a country church that I first made the acquaintance of that imperial star. It seemed to me more brilliantly beautiful than any of the dazzling gifts that had hung so magically on the illuminated tree. Its splendor is unearthly, putting diamonds and sapphires to shame. How people can live and be happy without ever gazing at such an object surpasses the understanding of any one who has once beheld and yielded to its charm. The splendors of Aladdin's Cave are for children, and fade in the light of advancing life, but these glories of the universe are for men and women, and grow brighter with the years.

The renown of Sirius is as ancient as the human race. There has never been a time or a people in which or by whom it was not worshipped, reverenced, and admired. To the builders of the Egyptian temples and pyramids it was an object as familiar as the sun itself. Its name is usually regarded as being derived from the Greek Σ   , the "Bright or Shining One," but it is also thought that it may be connected with Osiris. The familiar title of the "Dog Star" comes from its association with the *dies canicularia* of the Romans.

"As the movable Egyptian year," says George Cornewell Lewis, "was held to have originally begun at the heliacal rising of the Dog Star, which was contemporary with the ordinary commencement of the inundation of the Nile, this period was, by late writers, entitled the Canicular, or Sothiac, period, Sothis being the Egyptian name for the Dog Star."

Norman Lockyer identifies Sirius with the goddess Isis, or Hathor, who was personified by that star, and the temple of Isis at Dendera was, he avers, built to watch it. "It has been pointed out, times without number," he adds, "that the inscriptions indicate that by far the most important astronomical event in Egyptian history was the rising of the star Sirius at this precise time."

Sirius has sometimes been identified with the "Mazzaroth" of the Book of Job.

The great star is worthy of all its fame, not only by its magnificent beauty, but by the revelations which modern science has afforded us concerning it. While not comparable in actual luminosity with Rigel, Canopus, or even Arcturus, it immensely outshines the best of them to our eyes because of its relative nearness. Its distance is only about 50,000,000,000,000 miles (parallax 0 .37), so that it is really one of the nearest stars in the sky. Light requires about nine years to come to us from Sirius. Outshining the sun at least thirty times, it is so bright, even at that distance, that a special rank has been given to it in stellar photometry. Formerly all very bright stars were ranked as of the first magnitude, but greater exactness is now employed, the naked-eye stars being divided among eight magnitudes, running from 6 up to -1. Thus the faintest star visible to the naked eye is of magnitude 6; a star 2.51 times brighter is of magnitude 5; a star 2.51 times brighter than that is of magnitude 4, and so on up to magnitude

A star 2.51 times brighter than magnitude 1 is of magnitude 0; and one 2.51 times brighter than the 0 magnitude is of magnitude -1, a degree of brilliance which is attained by Sirius alone. In fact, Sirius exceeds magnitude -1, its real rank being -1.6. On the same scale the magnitude of the sun would be -26.3. The standard first magnitude s usually taken as being represented by the star Altair, although that star is not *exactly* of that magnitude. As a ready rule it may be said that each magnitude is two and a half times brighter than the next below it, and a difference of six magnitudes corresponds to an increase of one hundred times in brilliance. Sirius is about ten times as bright as Altair. While, if *seen from the same distance*, Sirius would appear at least thirty times as bright as the sun, at our actual distance from both the light received from the sun is to that received from Sirius in the ratio of about 7,000,000,000 to 1. While by no means the largest sun in the universe, Sirius is the largest sun in our part of space, and some indications have been detected that it may, to a certain extent, control the motion of the solar system. In other words, our sun and some of the nearer stars appear to form a group, or family, of which Sirius is probably the chief.

Sirius is an intensely white star, but its whiteness is shot with a tint of blue or green. It has not the purity of light of Spica. Owing also to its great brilliance, it twinkles incessantly, darting, in an unsteady atmosphere, rays of all the colors of the rainbow. The spectroscope shows that it is a sun at an earlier stage of development than ours. It is also a binary. A very massive companion, singularly faint for its size, revolves round it in a period of about fifty-three years. At present the distance between these stars is more than 6 . The small star is more than half as massive as Sirius, but ten thousand times less brilliant—one would say a dying sun linked by gravitation with another in the heyday of its life and splendor.

The constellation Canis Major, of which Sirius is the leader, is very striking in outline when well above the horizon. Some six degrees west of Sirius is seen the second star of the constellation, Beta ( ), or Murzim (Arabic *Al Murzim*, the "Announcer"), a name which Ideler says originated in the fact that this star rises ahead of Sirius, and thus appears to announce its coming. The remainder of the constellation should be viewed an hour or two later than that for which Chart IV is drawn, or a month later in the season, when it is farther from the horizon. It represents the hind-quarters of the imaginary dog. The star Epsilon ( ), or Adhara, perhaps the brightest in the group, is a double; colors orange and violet; distance 7 .5. The smaller star is of only the ninth magnitude. Delta ( ) is called Wezen, the "Weight," because "the star seems to rise with difficulty from the horizon," an excellent instance of the fanciful titles which the Arabs and others often gave to stars. Zeta ( ) is Furud, and Eta ( ) Aludra. The meaning of these names is uncertain. Allen says that the Arabs called Epsilon, Delta, Eta, and Omicron ( ) "The Virgins." But they had other names for them suggested by fancied resemblances as they rose sparkling from the desert.

From Canis Major the eye rises to Orion, the most glorious of all constellations:

"Whoso kens not him in cloudless night

Gleaming aloft, shall cast his eyes in vain

To find a brighter sign in all the heaven."

Brown, in his *Primitive Constellations*, undertakes to derive the name from the Akkadian Uru-anna, the "Light of Heaven." Whatever its origin, it is certainly very ancient. For some thousands of years it has been associated with a traditional giant who looms in the background of Greek mythology. In the classical atlases of the heavens Orion is represented as standing in an attitude of defiance, facing westward, brandishing a huge club above his head, and lifting his left arm, covered with a lion's hide, to meet the charge of Taurus, the "Bull." And under some such guise all mankind has seen him for untold ages—always a gigantic figure, always heroic in character, always defying or pursuing—the symbol of strength, courage, conquest, and victory. The same idea underlies every representation of this constellation; whether it be the mythical "Giant" of the East, or "Nimrod" or "Joshua" or the "Armed King" or the "Warrior" or the "Hunter," it is invariably the figure of a doer of great deeds which is presented to the imagination. And it must be said that the aspect of the constellation is in accord with such thoughts. No one can look at it without a stirring of the blood. It has something of the effect of a great battle-piece, and it is not surprising that they once endeavored in France to connect it with the name of Napoleon. Although its two chief stars are separated some eighteen degrees, and the central "Belt" forms a striking figure by itself, yet there is an unmistakable unity about the constellation, and one would hardly think of dividing it into separate groups. Singularly enough, this sense of oneness is borne out by the photographic discovery that a vast swirl of nebulous matter surrounds the entire constellation, and by the spectroscopic proof that nearly all of its stars belong to one type, which has become known as the "Orion type."

Perhaps the first feature of Orion that strikes the eye is the arrangement of the three nearly equal bright stars which form the Belt. Their Greek-letter names are Delta, Epsilon, and Zeta, and by these they are usually designated, but there is a great charm in their Arabic titles, which, in the same order, are *Mintaka*, "Belt"; *Alnilam* (from "String of Pearls"); and *Alnitah*, "Girdle." It will be observed that all of these names have a similar signification, and probably each of them was originally employed to designate the whole row.[2]

The Belt is remarkable in another way—it points very nearly toward Sirius; it is like a glittering signboard indicating the position of the brightest star in the sky. To hasty observation the row seems to be perfectly straight, although there is in reality a slight bend, and the distances separating the three stars appear to be exactly equal. The effect is as beautiful as it is surprising.

Below the Belt hangs a fainter row of stars constituting the "Sword." The central star of this row, Theta ( ), arrests the attention at once by a curious appearance of nebulosity, especially if it is examined with an opera-glass. A telescope shows it to be enveloped in one of the grandest nebulæ in the sky, the celebrated "Great Nebula of Orion." With a large glass its appearance is astonishing in the highest degree. Instead of being elongated like the great nebula in Andromeda, it is about as broad as long, with no single centre of condensation, but many curdled accumulations, interspersed with partial gaps, and a great variety of curved lines of brighter nebulosity, suggesting the misty skeleton of some nondescript monster impact of phosphorescent clouds. A large number of stars are scattered over or through it, and some of them seem clearly to be connected with it, as if created out of its substance. Unlike the Andromeda nebula, this shows only the spectrum of glowing gas, so that no such supposition as has been made in the other case—*viz.*, that it may be an outside universe—is admissible here. It is rather a chaos, rich with the elements from whose combinations spring suns and planets, and where the effects of organizing forces are just beginning to become manifest. It resembles a vast everglade filled with tangled vegetation and uncouth growths, but where the fertile soil, once cleared and drained, is capable of producing an enormous harvest.

On either side of the Belt, but far removed from it, shine the two great stars of Orion, Alpha ( ), or Betelgeuse (from an Arabic phrase meaning the "Armpit of the Central One"), and Beta ( ), or Rigel (from an Arabic phrase meaning the "Leg of the Giant"). These stars differ remarkably in color, Betelgeuse being orange-hued, and Rigel white. Although Betelgeuse takes precedence in the Greek-letter ranking, it is variable in brightness, sometimes exceeding Rigel in brilliance, and sometimes falling below it. The changes are uncertain in a long and as yet unascertained period. There is here an opportunity for an amateur to make valuable observations. But such observations must be continued over a considerable period of years.

Both stars are of immense actual magnitude. Their distance is so great that no trustworthy estimate of their parallax has yet been made. Rigel was put by Newcomb in his "XM" class, to which we have several times referred. It is without doubt one of the mightiest suns in the universe. It is also a double, and one of the finest in the sky. Close to its flaming rays the telescope reveals a small, intensely blue star. The distance is about 9 .5. In its general aspect Rigel resembles Vega, but the latter has a more decided blue tint. Scientific photometry gives the precedence in brightness to Vega, which is ranked as of magnitude 0.1, while Rigel is 0.3, which means that the first is one-tenth, and the second three-tenths of a magnitude below the 0 rank. It is very interesting to bring Rigel and Betelgeuse close together with a good sextant and then note the difference in their color.

The star Gamma ( ), or Bellatrix, the "Amazon" or "Female Warrior," marks the left shoulder of the imaginary giant. Astrological superstition connects this star with the fortunes of women. Kappa ( ), or Saiph, "Sword" (although it is far from the Sword), is in the right knee of the figure. The head is marked by a little triangular group of stars, the chief of which is Lambda ( ), a fine double, yellow and purplish; distance 4 .5. The "lion's hide" which Orion is represented as carrying on his left arm like a shield is shown by a bending row of small stars, beginning with Pi ( ) and running upward between Bellatrix and Aldebaran in Taurus. The reader who is not provided with a telescope is advised, at least, to employ an opera-glass in sweeping over the whole space included in Orion. It is a region superb in its beauty and grandeur. Around the Belt, particularly, the sky is filled with sparkling multitudes infinitely varied in size, color, and grouping. As already said, this part of the firmament contains an enormous spiral nebula, which, although it can only be seen in photographs, seems to manifest its presence to the eye by the significant arrangement of small stars in curving lines. A word should be added about the star Zeta, or Alnitah, at the southeastern end of the Belt. It is a triple, very remarkable for the indescribable color of its second largest component. The Russian astronomer Struve could find nothing exactly resembling it in tone in the whole gamut of spectral colors, and he invented a special name to describe it—*olivacea-sub-rubicunda*, which may be translated "ruddy-olive." It is 2 .5 from its larger companion. The third star is very faint, and distant 56 . When the telescope is directed to the star Sigma ( ) there comes into view an astonishing double group of stars, among which such colors as pale blue, "grape-red," ruddy, and "gray" have been detected. The effect upon the mind of seeing such combinations of tinted suns transcends all power of description. With the feeling of pleasure that they give goes a sense of staggering wonder.

West of Orion, beginning near Rigel, is seen the constellation Eridanus, the River Po. Its stars are interesting for their plainly streaming tendency rather than for their individual peculiarities. Rising slightly from the neighborhood of Rigel, the stream runs in a graceful curve under Taurus, and continues westward until it meets Cetus, where it turns downward toward the horizon, and then sweeps back eastward again, disappearing behind the southern horizon below Orion and Lepus. It has no large star visible in northern latitudes, but in the southern hemisphere it contains one of the brightest stars in the sky, Achernar, the "End of the River." All of the ancients saw a river in this part of the sky, a fact which does not surprise the observer when he has once noted the arrangement of the stars of Eridanus. Its stars are so numerous that the old uranographers seem to have grown weary of attaching letters to them; or rather, perhaps, the alphabet was too short to answer the demand, for no less than nine of them, beginning from the one thus lettered in Chart V, are called Tau ( ), as ¹, ², ³, etc. (For the origin of the association of Eridanus with the River Po, and with the story of Phaeton, see *Astronomy with theNaked Eye*).

The constellation Lepus, the Hare, below Orion, and marking the place where Eridanus turns finally to flow into the far south, is noteworthy only for its groupings of stars. It contains one star too faint to be seen with the naked eye near the western border of the constellation, below and to the right of the little group under Rigel, in Chart V, which is so intensely crimson that Hind likened its appearance to a *blood drop*.

We turn next to Taurus. On account of the beauty of Aldebaran and the Pleiades, this constellation hardly falls behind Orion in attractiveness. Aldebaran (Arabic *Al Dabaran*, the "Follower") is the chief star of the constellation and the leader of the group called the Hyades, a name which Lewis derives from the Greek word   , to rain, because their rising was connected with the beginning of the rainy season. Popularly the group is known as the "Letter A," whose form it imitates, although it is usually seen nearly upside down. The letter V would perhaps better represent our view of it. It is a glorious sight with an opera-glass. Aldebaran is distinctly red, but of a peculiar tone, which has frequently been called rose-red. Its redness is certainly unlike the orange tone of Betelgeuse. When gazing at it in a fanciful mood, I have often likened it imaginatively to an apple-blossom in color. Flammarion has translated the Hebrew name of this star, *Aleph*, as "God's Eye." Taurus, he says, is the most ancient of the signs of the zodiac, the first that the Precession of the Equinoxes placed at the head of the signs, and he adds that observational astronomy appears to have been founded at the epoch when the Vernal Equinox lay close to Aldebaran—*i. e.*, about three thousand years before the commencement of our era.

The beauty of Aldebaran, the singularity of the figure shaped by its attendants, the charming effect produced by the flocks of little stars, the Deltas and the Thetas, in the middle of the arms of the letter, and the richness of the stellar groundwork of the cluster, all combine to make the Hyades one of the most memorable objects in the sky; but no one can describe it, because the starry heavens cannot be put into words. Terrestrial analogies, and phrases applied to things seen on the earth, utterly fail to convey the impressions made by such spectacles. I can only again urge the reader to examine the Hyades with a good opera-glass on a clear night when there is no moonlight to interfere. Some one once said, "If you would test your appreciation of poetry, read Milton's *Lycidas*"; so I would say, If you would know how you are affected by nature's masterpieces in the sky, look at the Hyades.

The stars Theta ( ) and Sigma ( ) are both naked-eye doubles for sharp eyes. Try if you can see both of the pairs.

The Hyades represent the head of the imaginary bull, Aldebaran standing for the eye, while rows of stars running up toward Zeta ( ) and Beta ( ) figure the "golden horns." The Pleiades, the "Atlantid Nymphs," hang on the shoulder. They form a much more compact group than the Hyades, and possess no large star, their chief brilliant, Alcyone—Eta( )—being only of the third magnitude. But the effect of their combination is very striking and beautiful. In looking at them one can never refrain from quoting Tennyson's famous lines in which they are described as glittering "like a swarm of fireflies tangled in a silver braid." The adjective silvery exactly describes them. If you happen to glance at the sky at a point many degrees away from the place where they shine, your eye will inevitably be drawn to them. They have greater attractive power than a single large star, and the effect of their intermingled rays is truly fascinating. With an opera-glass they look like the glimmering candles on a Christmas-tree. Their mythological history and the many strange traditions pertaining to them I have described elsewhere, and shall not repeat here; but it should be said that there is not in all the sky any object comparable with the Pleiades in influence over the human imagination. The fancy of Maedler that Alcyone was the central sun of the universe, and the inference, so popular at one time, that it might be the very seat of the Almighty, have vanished in the limbo of baseless traditions; but the mystic charm of the Pleiades has been increased by the photographic discovery that they are involved in a wonderful mass of tangled nebulæ. Their distance is unknown, but evidently very great, some having put it at 250 light-years, corresponding to about 1,450,000,000,000,000 miles! If this is correct, Alcyone may be really one of the most gigantic suns in the universe. They appear to be travelling together like a flock of birds.

It is always an interesting question how many stars in the cluster can be seen with the naked eye. Many persons can detect only six, but better, or more trained, eyes see seven, or even nine. The telescope and photography reveal thousands thickly sprinkled over the space of sky that they occupy, or immediately around them. How many of these are actually connected with the group is unknown. One of the most persistent legends of antiquity is that of the "Lost Pleiad." Says Miss Clerke, in her *System of the Stars*:

"That they 'were seven who now are six' is asserted by almost all the nations of the earth from Japan to Nigritia, and variants of the classical story of the 'Lost Pleiad' are still repeated by sable legend-mongers in Victoria, by headhunters in Borneo, by fetish worshippers amid the mangrove swamps of the Gold Coast. An impression thus widely diffused must either have spread from a common source or originated in an obvious fact; and it is at least possible that the veiled face of the seventh Atlantid may typify a real loss of light in a prehistorically conspicuous star."

The name Pleiades is derived from the Greek   , to sail, because their heliacal rising occurred at the time when navigation opened in the seas of Greece, and their heliacal setting at the time of its close.

"... Rude winter comes

Just when the Pleiades begin to set."

But their religious significance seems always to have exceeded their practical importance as a sign of the seasons, and from the temples on the Acropolis of Athens to the sanctuaries of Mexico, Yucatan, and Peru they were regarded with reverence and awe. Modern popular fancies have been less reverential, and Alcyone and her attendants have been degraded to the figure of a "hen and her chickens." Our red-skinned predecessors on this continent were more poetical, for they saw in the Pleiades a group of lost children, and in old China they were starry sisters busy with their needlework.

High overhead, above Orion and Taurus, gleams Capella, the chief star of the constellation Auriga, the "Charioteer." This is also a white star, but no correct eye would confuse it with Rigel or Vega. It has none of the sapphire tint that is mingled in their rays, but is rather of the whiteness of cream. It is a very great star, not only in its apparent brilliance, but in actual luminosity. With a parallax of 0 .09, Newcomb calculated its luminosity at one hundred and twenty times that of the sun. It is a spectroscopic binary, the invisible companion revolving round it in a period of one hundred and four days. In spectroscopic character it closely resembles the sun, being in the same stage of development. Vogel's observations indicate that it is flying away from us at a speed of more than a million and a quarter miles per day; but, in contradiction to this, some have thought that it is increasing in brightness. A little elongated triangle of stars below and somewhat to the west of Capella serves to render its recognition certain to the beginner in star-gazing. In the evenings of early November, when one is in the northeast and the other in the northwest, it is interesting to compare Capella with Vega, both in brightness and in color. In late January evenings Capella is near the zenith for the middle latitudes of the United States, and at such times is a superb object.

The Milky Way pouring through Auriga increases the beauty of the spectacle.

The second star of Auriga, Beta ( ), or *Menkalina*, the "Shoulder," is also a spectroscopic binary with a period of only four days. It was the first binary of this class to be discovered. In 1889 Pickering found that its spectral lines were doubled every two days, from which he inferred the duplicate character of the star and calculated the period of revolution of its components.

Farther east we see Gemini, the "Twins." It is a very beautiful constellation, independently of the brightness of its leaders, Castor and Pollux, or Alpha ( ) and Beta ( ). The feet of the imaginary twins are dipped in the Milky Way nearly above the uplifted club of the giant Orion, and close to the summer solstice. The successive belts of stars crossing the figures of the Twins present an attractive appearance. Castor, although the literal leader of the constellation, is not now as bright as its neighbor, Pollux. A change of brightness must have taken place. Castor is a celebrated binary with a period of about one thousand years. The distance between the two stars composing it is about 5 .5, and, both being bright, they can be separated with small telescopes.

Pollux is very near the standard first magnitude in brightness. It has a slightly orange tint in contrast with the whiteness of Castor. Like Orion, Taurus, and Auriga, Gemini offers splendid fields of stars for the opera-glass. A cluster, M35, not far above the place of the summer solstice, is an object of rare beauty when seen with a low telescopic power.

South of Gemini shines the bright star Procyon in Canis Minor, the Lesser Dog. This star, whose name implies the "Preceder, or Announcer, of the Dog," because it rises a little ahead of Sirius, is the only bright star of its constellation. It is interesting for having a dusky companion whose existence was detected by the effects of its attraction before any telescope had revealed it. With this companion Procyon forms a binary system with a period of revolution of about forty years. The star Beta ( ) is named Gomeisa, from an Arabic word meaning the "Dim One." Procyon, Sirius, and Betelgeuse form a magnificent triangle, through which flows the Milky Way.

We now return to the western part of the sky, where we see, beyond Eridanus, the vast expanse covered by the constellation Cetus, the "Whale." The head lies on and over the equator above the western bend of Eridanus. It is marked by a striking group of stars, of which Alpha ( ), or Menkar, the "Nose," is the chief. The star Gamma ( ) is a fine double; colors yellow and blue; distance 2 .5. Below and toward the west will be found Omicron ( ), better known by its popular title of Mira, the "Wonderful." In some respects this is the most extraordinary of all variable stars. It excited great astonishment when its variations were first recorded in the seventeenth century. Most of the time it is entirely invisible to the naked eye; but once in about ten months it begins to brighten, and in a few weeks becomes conspicuous, sometimes equalling the second magnitude in brightness. Then it fades again, and in about three months disappears from naked-eye vision, although it is never lost to the telescope, which follows it down to the ninth magnitude, at which it remains, glowing redly, for several successive months. Its variations are more or less irregular both in period and in brightness. The causes are only conjectural. About all that we can say is that here is a sun which once every ten months blazes up to a thousand or fifteen hundred times its ordinary brilliancy. The imagination can work its will with such a star as that.

The western part of Cetus is marked by a striking group of stars shaped something like the bowl of an upturned dipper and by a lone, bright star still farther west, Beta ( ), or Deneb Kaitos, the "Tail of the Whale."

Above Cetus runs the long line of stars composing the constellation Pisces, now the leader of the zodiac, since it contains the Vernal Equinox. Alpha ( ), or Al Rischa, the "Cord," because it marks the ribbon imagined to bind two fishes together by their tails, is directly under the stars marking the head of Aries, to which we have already referred. It is a double of very singular colors—green and blue. The distance is about 3 .6. From Al Rischa the stars of the constellation stream northward to the figure of the Northern Fish, whose nose touches Andromeda, and westward to the Western Fish, which is situated under the Great Square of Pegasus. The extraordinary tendency of the stars of Pisces to run in streaming lines has been spoken of in Chapter III.

The other stars and constellations now visible are already familiar to us. But we turn again for a moment to Polaris, which, being practically fixed in the sky, can be seen at any season. I have referred to the fact that this star for a long series of centuries has been a universal guide to all the inhabitants of the northern hemisphere. In that character its history is no less romantic than practically important. One of the deepest impressions of my childhood was produced by an acquaintance with a remarkable man who at that time seemed to me to be a most wonderful traveller, since he had seen the Gulf of Mexico, the Everglades of Florida, the Dismal Swamp of Virginia, and, according to his story (which no boy would doubt), had battled with alligators and tasted the delights of vagabond life on the great cotton plantations of the South. I think he was the first who ever pointed out the North Star to me, and he fired my imagination by tales of its connection with the escape of negro slaves—escapes in which he professed to have played a part. Many long winter evenings he sat by my father's fireside and fascinated his hearers with narratives of his adventures. But nothing interested me more than what he said of the slaves following the lead of the North Star, through the darkness of tangled swamps, among deadly moccasins and lurking alligators, always fixing their eyes upon "the star," falling on their knees to it as their only friend and guide. Trembling at the bay of pursuing bloodhounds, they would lie in concealment during the daylight hours, and as soon as night came on would look for their celestial sentinel, and follow unquestioningly its indication of the way to freedom. However apocryphal these stories may have been, they certainly had a basis of truth, and the impressions then produced upon my mind concerning the character of Polaris as the sure friend of those who are lost and in trouble have remained undimmed in my memory. What a triumph will be that of the man who first visits the north pole by night, and sees that star gleaming directly over his head, while all the constellations solemnly circle about it, unresting and unsetting!

FOOTNOTES:

[2] It should be said that throughout this book I am indebted for many of the translations of star names to Richard Hinckley Allen's *Star Names and Their Meanings*, the most complete work of its kind in existence.

V THE PLANETS

The beginner will often be troubled in his observations by the presence in some constellation of a brilliant object which outshines all of the stars shown in his charts, and is plainly an interloper among them. He may at once set the stranger down for one of the planets—it may be Jupiter, Saturn, Mars, or Venus, or possibly, if close to the horizon, Mercury. Uranus and Neptune will not disturb his equanimity, for the latter is never, and the former seldom, visible to the naked eye.

Practice will quickly enable him to distinguish a planet from the true stars, both by its greater apparent size and by the quality of its light. The planets do not twinkle as do the stars. This arises from the fact that they present measurable disks which reflect the sunlight, but do not shine with a light of their own. No star shows a real disk, even when viewed with a powerful telescope. The stars are mere points, and the larger and better the telescope the smaller they appear. This is not to say that they do not look brighter in a telescope, for the larger stars are dazzling when viewed with a glass of large aperture; but they are so distant that the mightiest of telescopes cannot reveal their real surfaces in the form of disks. The apparent disks which they present are due entirely to irradiation, and the higher the power the smaller these spurious disks appear.

Another way in which the beginner may identify a planet is by observing its motion. No planet remains long in the same position with regard to neighboring stars. They all travel, at varying rates, from west to east through the sky. But this motion is not constant, and at times it is reversed. In the cases of Mars, Jupiter, and Saturn the reversal is due to the fact that when they are in opposition to the sun the earth, being nearer the sun than they are, outfoots them in eastward motion, so that they appear for a time to move backward on their orbits. It is like a fast train passing a slow one on a parallel track; to an observer on the fast train the slow one seems to be either standing still or moving backward. But Mercury and Venus, being nearer the sun than the earth is, have at times a backward motion which is real. Let us consider them only when they appear as "evening stars." From "superior conjunction" (i. e., the point occupied by the planet when it is on the opposite side of the sun from the earth) to "greatest eastern elongation" (greatest apparent distance from the sun in the evening sky) both Mercury and Venus move eastward among the stars; from "greatest eastern elongation" to "inferior conjunction" (*i. e.*, the point occupied by the planet when it is between the earth and the sun) they move westward among the stars, or, in other words, approach the sun.

The motions of Mercury and Venus are comparatively swift, particularly that of the former. Few persons have ever seen Mercury, because of its nearness to the sun. When well seen it is brighter than any first-magnitude star. As an "evening star" it appears in the west immediately after sunset about once every four months (more precisely once every 116 days). It remains within view about twenty days, but can be easily distinguished only for a week or so when it is nearest eastern elongation. Every almanac gives the dates of its appearances.

Venus, being farther from the sun, travels less rapidly. It reappears in the evening sky once in every 584 days, gradually withdrawing from the sun, and growing brighter until it reaches greatest eastern elongation, which may be as much as forty-seven degrees from the sun, after which it approaches the sun, still becoming brighter for several weeks, until at last it is lost in the glare of the sunlight. During its excursions in the evening sky (and the same is true of its morning apparitions), Venus becomes the most brilliant object in the starry heavens, so brilliant, in fact, that many persons can hardly be persuaded that it is not an artificial light, or some extraordinary phenomenon in space. In the telescope it shows (as does Mercury, also) phases like those of the moon, and when it is seen in the form of a narrow crescent it becomes one of the most charming objects imaginable. For more details about Mercury, Venus, and the other planets, the reader may consult *Astronomy with the Naked Eye.*

Mars, Jupiter, and Saturn are more likely to cause confusion to the beginner by getting "mixed up" with the stars of the constellations he is studying, because they travel all round the sky, and may appear in turn in each of the zodiacal constellations at any hour of the night. The zodiacal constellations are twelve in number—Aries, Taurus, Gemini, Cancer, Leo, Virgo, Libra, Scorpio, Sagittarius, Capricornus, Aquarius, and Pisces—and they lie in succession along the course of the ecliptic.

Mars is not remarkably brilliant except when it is in opposition to the sun, which happens once every 780 days; but some of the oppositions are much more important than the average, because they occur when Mars is relatively near the earth. This planet is always distinguishable by its ruddy color. In case it is mistaken for a star, the error can be corrected by watching it for a few successive nights, when its motion will become clearly apparent. On the average it moves eastward about half a degree per day.

Jupiter, always very conspicuous when in view, outshines even Sirius, though lacking the scintillation characteristic of that great star. Its light has a slightly yellowish tint, and is remarkably steady. Since it requires nearly twelve years to make a revolution round the sky, Jupiter's motion is not immediately apparent. It remains for a long time in any constellation in which it may be found, travelling eastward, on the average, about 5 of arc, or one-sixth of the apparent diameter of the moon, per day. In a month it moves about two and a half degrees.

Saturn is yet more deliberate in its movements. Requiring almost thirty years for a revolution, it may remain more than two years in the same constellation, and its real motion will only become evident upon careful observation continued for several weeks.

The best way to recognize the planets with certainty is to look up their positions with the aid of the *American Ephemeris and NauticalAlmanac*, published annually by the Government at Washington. There the right ascensions and declinations of all the planets are given for any time of the year. Having these, you may find on the large-scale charts the approximate place of the planet sought, and, if you choose, indicate its position with a pencil-mark.

The study of the planets, even without telescopic aid, has a charm hardly less potent than that of the stars. Mercury is fascinating because of the difficulty of seeing him in the light of twilight or dawn. The ancients were greatly puzzled by his dodges, and some of them thought that he was a double personality, and gave him two names, one for his morning and the other for his evening apparitions. With the Egyptians he was respectively Set and Horus, and with the Greeks Apollo and Hermes. The same was true of Venus, who was Phosphorus in the morning and Hesperus in the evening.

Venus, after she passes the half-moon phase, becomes so bright that she simply overpowers all stars in her neighborhood. Her splendor seems almost supernatural, and she has frequently been seen at high noon, a point of intense light burning in the blue sky.

Jupiter's entrance into any constellation immediately alters its familiar aspect, and he becomes its unquestioned leader, and remains such until his slow eastward motion carries him on to reign in another quarter of the firmament. He is never more impressive than when, in consequence of the annual revolution of the heavens, he rises late some night and takes the lingering star-gazer by surprise. Then all the stellar hosts that for hours have held the watcher spellbound cease their incantation in the presence of this great counter-charmer, to whose power they, too, seem to bow. Although Venus at her brightest outshines Jupiter, she lacks a certain majesty which he alone possesses. His light is calm, steady, insistent, commanding. He does not look like a star, but rather a *superstar*. If he beams at all, it is not the hurried scintillation of the twinkling multitude around him. Rising through a moisture-laden and wind-swept sky, where the stars are like pulsating atoms, shaken apart and scattered in tinsel showers of rainbow sparks, he glows unflickering, recognizing the aerial tumult only by a deepening of color which makes him the more imposing. As he mounts the heights of the sky he gleams ever brighter and ever steadier, and, casting off the tarnish of the horizon, his supereminent light glows with a splendor that is amazing. If you have an eye that can detect one or two of Jupiter's moons hiding close in his rays, you may boast of your powers of vision, for that feat has been accomplished by very few human beings. Humboldt heard of a German "master tailor" who could do it. There are a few other cases on record. Most persons cannot see them even with the aid of a strong opera-glass. There is a superstition that they can be seen with a looking-glass, but it is only ghostly reflections that are thus perceived—perhaps as real as any other ghosts.

Saturn, although as bright as a first-magnitude star, is somewhat disappointing as a naked-eye object, owing to the relative dulness of its light. Like Jupiter, it shines with great steadiness, and a practised eye could not mistake it for a fixed star. But its appearance without a telescope gives no hint of the unearthly beauty with which it astonishes the beholder when its rings are rendered visible. Not to have seen those rings at least once in a lifetime, as they appear in a powerful telescope, is to have missed one of the supreme spectacles of creation.

Mars is never very brilliant except during favorable oppositions, when, approaching within less than 40,000,000 miles of the earth, it hangs in the midnight sky, gleaming red like a portent of disaster. The aspect of Mars at such times is truly alarming. It is surprising to see what a quantity of stained sunlight a world only about four thousand miles in diameter is able to reflect across so vast a gap of space. The reason why the ancients connected Mars with the god of war is plain enough when he puts on his color.

Close conjunctions of the bright planets are exceedingly interesting phenomena. Mars and Jupiter seen together when the former is near one of its favorable oppositions make a scene of strange beauty. After long intervals of time several of these great planets sometimes assemble in the same quarter, and such conjunctions are always memorable occurrences. The stars are forgotten in the presence of this new constellation, and yet the tiniest of the sparks that seems to hide its light in the depths beyond would master these great planets and make gravitational slaves of them, as the sun does.

The planets are so conspicuous to our eyes, because of their relative nearness, that it is not easy for the beginner in such studies to realize how insignificant they actually are. But suppose that one could fly like a spirit away from the earth and the neighborhood of the sun, out into the deeps of interstellar space. As he moved away the planets would seem to be swallowed up, one after the other, in the solar rays. First Mercury would disappear, as if it had fallen into the sun. It would be just like two neighboring lights which appear to draw together and blend into one as the observer travels away from them, the greater swallowing the less. Then brilliant Venus would go, plunging into the great solar furnace, to be seen no more. Next the earth would follow in the perspective holocaust. Mars would seem to draw nearer until he, too, disappeared; Jupiter would follow; then Saturn; then Uranus, and finally Neptune. When the last planet was gone the sun would be seen shining alone, unattended, as if he had never had any planets. Thus it may be with the stars; most of them may have systems of planets circling round them, but at our distance these planets are concealed in the rays of their primaries.

One would not need to go so far away as the stars in order to see the sun apparently swallow his planets, as Saturn was fabled to have swallowed his children. But as one approached the stellar region, the sun itself would become a mere star. Fainter and fainter it appears, glimmering and twinkling, deprived of its dominance, stripped of its splendor, a pitiful spark now instead of an all-ruling and blinding maker of daylight, until at last the far voyager from the earth, gazing with his soul in his eyes, straining his vision to the utmost to hold that glinting point clear of its fellows, *for it is his sun,* suddenly, as a momentary film blurs his sight, loses it, and henceforth seek as he may among the countless hosts that spangle the firmament, he will never again find the day-star under whose cheery beams he was born! Hidden in the Milky Way, one would have no more chance of recognizing the sun than of finding a particular grain of sand on the sea-shore. Man physical is as insignificant as the rock he dwells on and as the eye-searing orb that lights him at his daily work; but man spiritual is as great as the universe—and greater!

APPENDIX
URANOGRAPHY OR HEAVENLY DESCRIPTION OF THE CHURCHMEN

Many readers may be interested in seeing a list of the names given to the constellations when, as mentioned in the Introduction, the starry sky was "Christianized." In the seventeenth century Julius Schillerius put forth his *Cœlum Stellatum Christianum*, and Jacobus Bartschius a celestial globe, in which all of the well-known constellations received new and strictly orthodox names. Unfortunately the sponsors for these names did not always agree in their choice, and a certain Harsdorfius (who may have been the poet Philip Harsdoerfer, born at Nuremberg in 1607) added to the confusion by further varying the selection. Wilhelm Schickard also introduced variations. In the following list the first of the "Christian" names given is that chosen by Schillerius, while their variants are due to either Harsdorfius, Schickard, or Bartschius:

Aries—St. Peter—Abraham's Ram.

Taurus—St. Andrew—The Burnt Sacrifice.

Gemini—St. James the Elder—Jacob and Esau.

Cancer—St. John the Evangelist.

Leo—St. Thomas—The Lion of Judah. (Observe that the variants are generally more imaginative.)

Virgo—St. James the Younger—The Virgin Mary.

Libra—St. Philip—Belshazzar's Balances.

Scorpio—St. Bartholomew.

Sagittarius—St. Matthew—Ishmael.

Capricornus—St. Simon.

Aquarius—St. Jude—Naaman.

Pisces—St. Mathias—The Gospel Fishes.

Ursa Minor—St. Michael—One of Elisha's Bears—The Wagon of Joseph.

Ursa Major—St. Peter's Fishing-boat—Elisha's other Bear—The Chariot of Elias.

Draco—The Innocents—The Dragon Infernal. (Quite a difference of opinion.)

Boötes—St. Sylvester—Nimrod.

Coma Berenices—The Scourge of Christ—Absalom's Hair—Samson's Hair.

Corona Borealis—The Crown of Thorns—Queen Esther's Crown.

Hercules—The Three Wise Men of the East—Samson.

Lyra—The Saviour's Manger—David's Harp.

Cygnus—The Cross of Calvary.

Cassiopeia—St. Mary Magdalen—Bathsheba.

Cepheus—St. Stephen—Solomon. (Solomon seems a better choice.)

Perseus with Medusa's Head—David with the Head of Goliath—St. Paul.

Andromeda—The Holy Sepulchre—Abigail. (The last reverses Andromeda's romance with a vengeance.)

Auriga—Jacob—St. Jerome.

Ophiuchus et Serpens—St. Benedict—St. Paul and the Viper. (The latter very pat.)

Sagitta—The Lance of Calvary—Jonathan's Arrow.

Aquila—St. Katharine—The Standard of Rome.

Delphinus—The Canaanitish Woman's Pitcher—Leviathan.

Equuleus—The Mystic Rose.

Pegasus—St. Gabriel—Jeremiah's King of Babylon.

Triangulum—St. Peter's Mitre—Emblem of the Trinity.

Cetus—Sts. Joachim and Anna—Jonah's Whale.

Eridanus—The Red Sea with Moses Crossing It—The Brook of Cedron.

Orion—St. Joseph—Joshua. (The last a good choice.)

Lepus—Gideon's Fleece.

Canis Major—Tobias's Dog—St. David.

Canis Minor—The Paschal Lamb.

Argo Navis—Noah's Ark. (Inevitable!)

Hydra—The River Jordan.

Crater (together with Corvus)—The Ark of the Covenant.

Corvus (according to Schickard)—Elias's Crow.

Centaurus—Abraham and Isaac.

Lupus—Jacob.

Ara—The Altar of Incense.

Corona Australis—David's Crown—Solomon's Crown.

Piscis Austrinus—The Widow's Meal Barrel—St. Peter's Fish with Money in Its Mouth.

Grus

Phœnix}—Aaron.

Indus

Pavo   }—Job.

Apus

Chameleon   }—Eve.

Piscis Volans

Triangulum Australe—The Cross of Christ. (At that time the Southern Cross seems not to have been known.)

Dorado

Toucan   }—St. Raphael.

Hydrus

The southern constellations, Grus, Phœnix, Indus, Pavo, Apus, Chameleon, Piscis Volans, Triangulum Australe, Dorado, Toucan, and Hydrus, were all named by Bayer at the beginning of the seventeenth century, so that the revisers were not upsetting any antique legends in giving them more sacred names.

LETTERS OF THE GREEK ALPHABET EMPLOYED IN URANOGRAPHY

—Alpha

—Beta

—Gamma

—Delta

—Epsilon

—Zeta

—Eta

—Theta

—Iota

—Kappa

—Lambda

—Mu

—Nu

—Xi

—Omicron

—Pi

—Rho

—Sigma

—Tau

—Upsilon

—Phi

—Chi

—Psi

—Omega

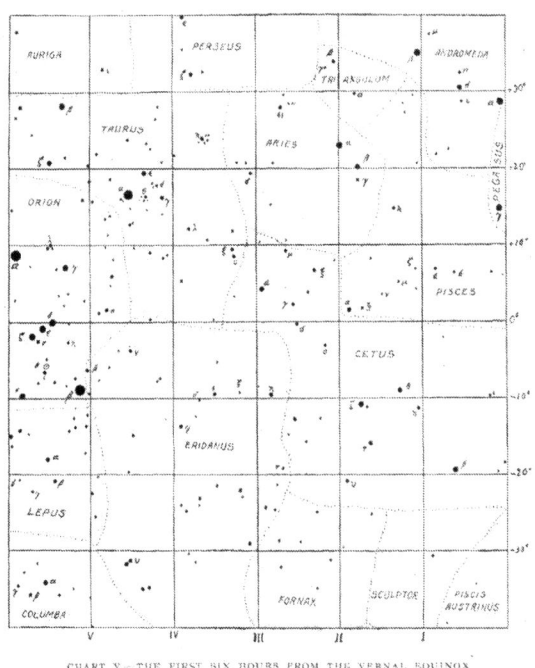

CHART V—THE FIRST SIX HOURS FROM THE VERNAL EQUINOX

CHART V—THE FIRST SIX HOURS FROM THE VERNAL EQUINOX

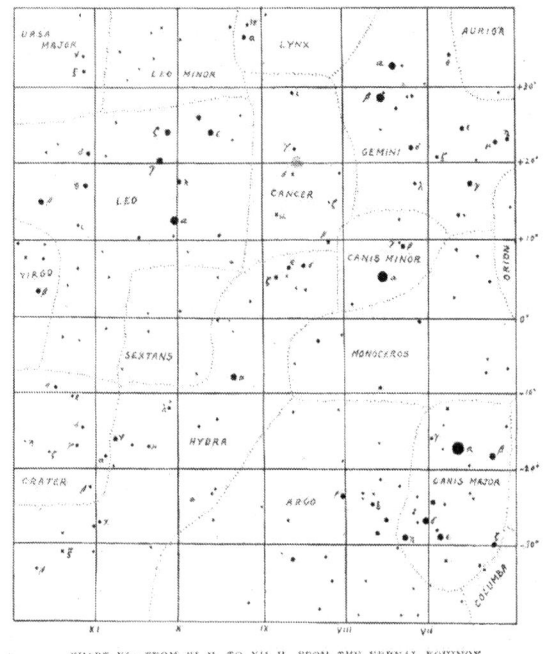

CHART VI—FROM VI H. TO XII H. FROM THE VERNAL EQUINOX

CHART VI—FROM VI H. TO XII H. FROM THE VERNAL EQUINOX

68

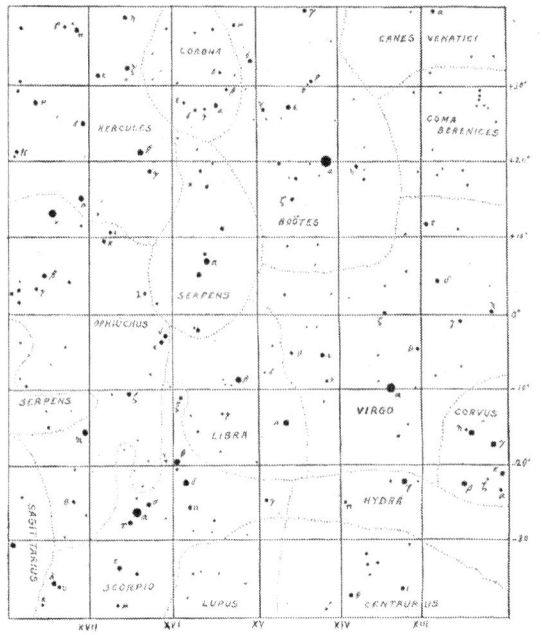

CHART VII—FROM XII H. TO XVIII H. FROM THE VERNAL EQUINOX

CHART VII—FROM XII H. TO XVIII H. FROM THE VERNAL
EQUINOX

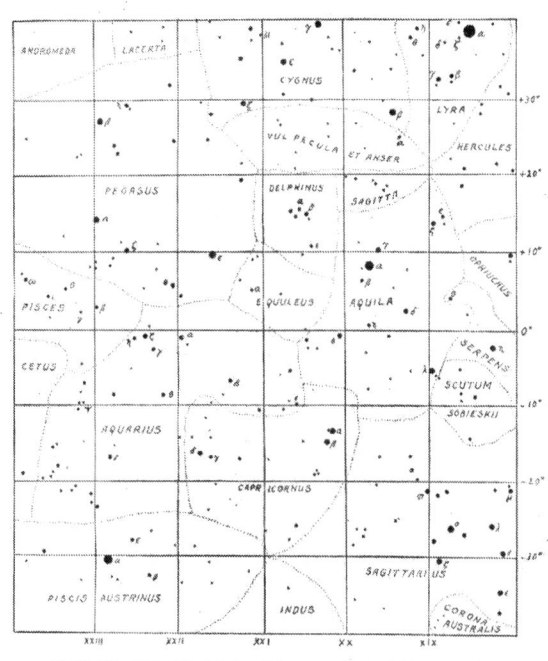

CHART VIII—FROM XVIII H. TO XXIV H. FROM THE VERNAL EQUINOX

CHART VIII—FROM XVIII H. TO XXIV H. FROM THE VERNAL
EQUINOX

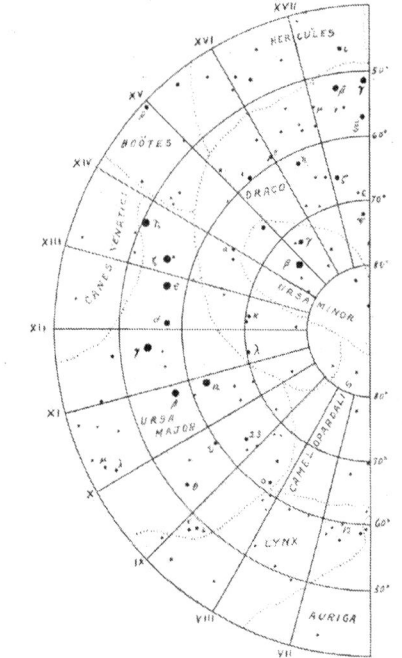

CHART IX—POLAR CONSTELLATIONS FROM VI H. TO XVIII H.

CHART IX—POLAR CONSTELLATIONS FROM VI H. TO XVIII H.

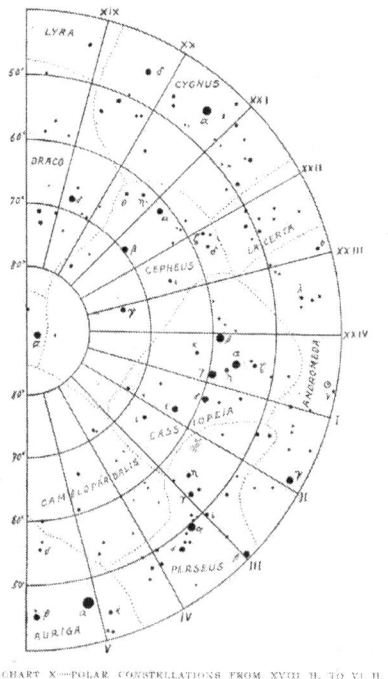

CHART X—POLAR CONSTELLATIONS FROM XVIII H. TO VI H.

CHART X—POLAR CONSTELLATIONS FROM XVIII H. TO VI H.

PRONUNCIATION OF STAR AND CONSTELLATION NAMES

Achernar (ä-kĕr-när)

Albireo (ăl-bĭ-rē-ō)

Alcyone (ăl-sĭ-ŏ-nē)

Aldebaran (ăl-dĕb-àr-ăn)

Algenib (ăl-ḡén-ib)

Algenubi (ál-ḡen-ŭ-bĭ)

Algieba (ăl-jĕ-bà)

Algol (ăl-gol)

Algorab (ál-go-rắb)

Alioth (ál-ĭ-öth)

Alkalurops (ál-kā-lŭ-rŏps)

Alnılam (ál-nĭ-lăm)

Alnitah (ăl-nĭ-tắh)

Almaack (ál-mā-ắck)

Alphacca (ăl-fắk-kà)

Alphard (ăl-fắrd)

Alpheratz (ăl-fé-rătz)

Alrischa (ăl-rĭ-sħà)

Alrucaba (ăl-rŭ-cắ-bà)

Altair (ăl-tắr or ăl-ĭă-ĭr)

Aludra (ă-lŭ-dﭏä)

Andromeda (ăn-dﭏŏm-ē-dà)

Antares (ăn-tắ-rēz)

Antinous (ăn-tĭn-ŏ-ŭs)

Aquarius (à-kwā-rĭ-ŭs)

Aquila (ắk-wĭ-là)

Arcturus (ärk-tŭ-rŭs)

Argo Navis (är-gō ńà-vĭs)

Aries (á-rēz or ắ-rĭ-ēs)

Auriga (äw-rĭ-ḡà)

Baten Kaitos (bắ-tĕn kītŏs)

Bellatrix (bĕl-lắ-trĭx)

Benetnasch (bē-nĕt-născh)

Betelgeuse (bĕt-ĕl-ḡooz or bĕt-ĕl-gēz)

Boötes (bb̄-ŏ-tēz)

Camelopardalis (căm-ĕĺ-ō-pắ´r-dā-lĭs)

Canes Venatici (cắ-nēz vĕn-ắt-ĭ-sī)

Canis Major (cắ-nĭs mắjor)

Canis Minor (cắ-nĭs mίńor)

Canopus (càn-ŏ-pus)

Capella (cā-pĕl-là)

Caph (kāff)

Capricornus (cắp-rī-kór-nus)

Cassiopeia (căs-sĭ-ō-pĕ́-yà)

Centaurus (cĕn-táw-rus)

Cepheus (śē-fē-us or śē-fūs)

Cetus (śē-tŭs)

Coma Berenices (cŏmā bēr-ĕ-nĭ́-sēs)

Corona Borealis (cŏ-rŏ́-nà bō-rē-ắ-lis)

Corvus (côŕ-vus)

Crater (crắ-ter)

Cygnus (sĭĝ-nus)

Delphinus (del-fί-nus)

Deneb (dĕ́ń-eb)

Denebola (dē-nĕb-ō-là)

Draco (dŕā-co)

Dubhe (dūb̄-hĕ̆)

Eltanin (ĕĺ-tà-nĭn)

Equuleus (ē-kwóo-lē-ŭs)

Eridanus (ē-rĭ́đ-ā-nus)

Fomalhaut (fŏ́-măl-hôt)

Fornax (fôŕ-naks)

Gemini (jĕm-ĭ-nī)

Giedi (jĕ́-dĭ)

Gienah (jĕ́-nah)

Gomelza (gō-mĕl-zȧ)

Hamal (hăm-al)

Hercules (hĕr-kū-lēz)

Hyades (hĭ-ȧ-dēz)

Hydra (hĭ́-drȧ)

Lacerta (lȧ-sĕr-ta)

Leo (lĕ́-ō)

Lepus (lé-pus)

Libra (lĭ́-brȧ)

Lyra (lĭ́-rȧ)

Maia (mắ-yȧ)

Marfak (már-făk)

Markab (már-kăb)

Megrez (mĕ́-grĕz)

Menkab (mĕn-kăb)

Menkalina (mĕn-kȧ-lĭ-náh)

Merope (mĕr-ō-pē)

Mesarthim (mē-sār-thĭm)

Mintaka (mĭń-tȧ-kȧ)

Mira (mĭ́-rȧ)

Mirach (mĭ́-rak)

Mizar (mĭ́-zȧr)

Monoceros (mō-nŏ́ś-ĕr-ŏs)

Murzim (mŭ́r-zĭm)

Ophiuchus (ŏ-fĭ́-ŭ-kus)

Orion (ō-rĭ́-ŏn)

Pegasus (pĕ́ǵ-ā-sŭs)

Perseus (pĕ́r-sē-ŭs or pĕ́r-sūs)

Pisces (pĭ́ś-sēz)

Piscis Austrinus (pĭ́s-sĭs aus-trĭ̄-nus)

Pleiades (plĕ́-ăd-ēz or plĭ-ăd-ēz)

Polaris (pō-lȧŕ-ĭs)

Pollux (pŏl-lux)

Porrima (pŏ́r-rĭ-mȧ)

Præsepe (prē-sĕ́-pē)

Procyon (prō-sĭ́-ŏn)

Ras Algethi (rȧs ăĺ-gĕ̆-thĭ)

Rastaban (rȧs-tà-bāń)

Regulus (rĕḡ-ū-lús)

Rigel (rĭ-ḡĕl or rĭ-jĕl)

Sagitta (sȧ-jĭt-tȧ)

Sagittarius (sȧ-jĭt-tȧ-rĭ-ús)

Scheat (she-ăt)

Schedar (shĕđ-där)

Scorpio (skór-pĭ-ō)

Scutum Sobieskii (skū-tŭm sö-bĭ-ĕś-kĭ-ī)

Serpens (sĕr-pens)

Sirius (sĭŕ-ĭ-ús)

Spica (spĭ-kȧ)

Taurus (tāú-rŭs)

Thuban (thu-bäń)

Triangulum (trī-ăń-ġū-lŭm)

Ursa Major (ûŕ-sȧ mắ-jor)

Ursa Minor (ûŕ-sȧ mĭ-nor)

Vega (vḗ-ḡȧ)

Vindemiatrix (vĭn-dḗ-mĭ-ắ-trĭx)

Virgo (vĕŕ-go)

Vulpecula (vŭl-pĕk-ū-lȧ)

Wesen (wắ-zĕn)

Zavijava (zȧ-vĭ-jắ-vāh)

Zubenelgenubi (zū-bĕn-ĕl-jen-ū-bĭ)

Zubeneschemali (zū-bĕn-ĕs-she-mắ-lĭ)

INDEX

"A," the letter, 108.

Achernar, 107.

Adhara, 100.

Afternoon of the year, 72.

Albireo, 64.

Al Chiba, 33.

Alcor, 41.

Alcyone, 110.

Aldebaran, 107 *et seq.*

Alderamin, 91.

Aleph, 108.

Alfirk, 92.

Algedi, 78.

Algenib, 86.

Algieba, 36.

Algol, 86.

Algorab, 33.

Al Hiba, 33.

Alioth, 41.

Allen, R. H., quoted, 41, 57, 64, 70, 78, 101, 102.

Almaack, 82.

Alnilam, 102.

Alnitah, 102.

Alpha Andromedæ, 12, 81.

Aquarii, 77.

Aquilæ, 65.

Arietis, 83.

Aurigæ, 111.

Boötes, 67.

Canis Majoris, 94.

Canis Minoris, 113.

Capricorni, 78.

Cassiopeiæ, 88.

Centauri, 65.

Cephei, 91.

Ceti, 114.

Corvi, 33.

Cygni, 63.

Draconis, 45, 48.

Geminorum, 113.

Herculis, 59.

Hydræ, 35.

Leonis, 35.

Libræ, 70.

Lyræ, 48, 61.

Orionis, 104.

Pegasi, 81.

Piscis Austrinus, 75.

Piscium, 115.

Scorpii, 54.

Serpentis, 57.

Tauri, 107.

Ursæ Majoris, 40.

Ursæ Minoris, 46.

Virginis, 29.

Alphard, 34.

Alpheratz, 12, 81.

Al Rischa, 115.

Altair, 65.

Aludra, 100.

American Ephemeris, 122.

Andromedæ, 12, 81.

Antares, 54.

Antinous, 66.

Aonian Dragon, 34.

Aquarius, 77 *et seq.*

Aquila, 65.

Aratus, quoted, 30.

Arctophilax, 68.

Arcturus, 43, 67 *et seq.*

Argonautic Expedition, 34.

Aries, 83, 84.

Auriga, 111.

Autumnal Equinox, 71.

Barnard, E. E., quoted, 59.

Beehive, the, 38.

Bell, Dr. Louis, quoted, 90.

Bellatrix, 105.

Belt of Orion, 102, 105.

Beta Andromedæ, 81.

Aquarii, 78.

Arietis, 84.

Aurigæ, 112.

Canis Majoris, 100.

Canis Minoris, 113.

Capricorni, 78.

Cassiopeiæ, 88.

Cephei, 92.

Ceti, 115.

Corvi, 33.

Cygni, 64.

Geminorum, 113.

Leonis, 37.

Libræ, 70.

Lyræ, 63.

Orionis, 104.

Pegasi, 81.

Scorpii, 55.

Tauri, 109.

Ursæ Majoris, 40.

Ursæ Minoris, 46.

Virginis, 32.

Betelgeuse, 104.

Boötes, 43, 69.

Cancer, 38.

Canes Venatici, 42, 69.

Canis Major, 100.

Canis Minor, 113.

Capella, 43, 111.

Caph, 88.

Capricornus, 78.

Carman, Bliss, quoted, 22.

Cassiopeia, 47, 88.

Castor, 113.

Cepheus, 91.

Cetus, 114.

Chair, Cassiopeia's, 87.

Cheops, pyramid of, 45.

Christianization of sky, 12.

Clerke, Agnes M., quoted, 111.

Coma Berenices, 38.

Constellations, their uses, 12, 13.

Cor Caroli, 42.

Cor Hydræ, 34.

Corona Borealis, 66.

Corvus, 32.

Crater, 33.

Crete, discoveries in, 62.

Cupid's Arrow, 80.

Cygnus, 63.

61 Cygni, 64.

Dabih, 78.

Dana, R. H., quoted, 77.

Delphinus, 79.

Delta Andromedæ, 81.

Canis Majoris, 100.

Cassiopeiæ, 47, 88.

Corvi, 33.

Orionis, 102.

Sagittarii, 56.

Serpentis, 57.

Ursæ Majoris, 40.

Virginis, 32.

Demon star, 86.

Deneb, 63.

Deneb Kaitos, 115.

Denebola, 37 *et seq.*

Dipper in Sagittarius, 56.

Dippers, the, 29.

Discipline in the sky, 9.

Donati's comet, 69.

Draco, 44 *et seq.*

Dream of the universe, 91.

Dubhe, 41.

Durchmusterungs, 12.

Eltanin, 46.

Emerson, R. W., quoted, 22, 53, 61.

Epsilon Canis Majoris, 100.

Boötes, 69.

Lyræ, 63.

Orionis, 102.

Pegasi, 81.

Serpentis, 57.

Ursæ Majoris, 40.

Virginis, 32.

Equator, 13.

Equinoctial colure, 88.

Equinoctial storms, 71.

Equinoxes, 13.

Eridanus, 106.

Esculapius, 57.

Eta Cassiopeiæ, 89.

Canis Majoris, 100.

Eta Tauri, 107.

Expectancy of astronomers, 27.

Feet of Ursa Major, 42.

Field of the nebulæ, 32.

Flammarion, quoted, 108.

Fomalhaut, 75.

Furud, 100.

Galileo, 38.

Gamma Andromedæ, 82.

Aquarii, 78.

Arietis, 84.

Capricorni, 79.

Ceti, 114.

Corvi, 33.

Draconis, 46.

Leonis, 36.

Lyræ, 63.

Orionis, 105.

Pegasi, 80.

Virginis, 32.

Gardens of the sky, 72.

Garnet star, 92.

Gateway of souls, 39.

Gemini, 113.

Gemma, 66.

Gienah, 33.

"God's Eye," 108.

Gomeisa, 114.

Great Bear, 40.

Great Dipper, 40.

Great Square of Pegasus, 80.

Great Year, Plato's, 62.

Greenwich of the sky, 14.

Halley's comet, 65.

Hamal, 83.

Hathor, temple of, 46.

Hercules, 59.

Hesperus, 123.

Hexagon of Orion, 28.

Hole in the sky, 59.

Horus, 122.

Hyades, 108.

Hydra, 34.

Influence of the stars, 10, 18, 22, 27, 29, 35, 39, 44, 53, 68, 75, 93, 102, 110, 116.

Isis, 97.

Jason, 34.

Job's Coffin, 79.

Job's Star, 68.

Jupiter, 121, 123.

Kappa Cassiopeiæ, 90.

Orionis, 105.

Karnak, 46.

Kochab, 49.

Laconian Key, 87.

Lambda Ophiuchi, 57.

Orionis, 105.

Language for celestial marvels, 61.

Learning the stars, ease of, 17; best season for, 93.

Leo, 35.

Lepus, 107.

Lewis, G. C., quoted, 25, 97.

Libra, 70.

Lockyer, Norman, quoted, 97.

Longfellow, H. W., quoted, 71.

Lucky stars, 77, 78.

Lyra, 61.

Magellan Clouds, R. H. Dana on, 77.

Marfik, 58.

Markab, 81.

Mars, 121, 124.

Mazzaroth, 98.

Medusa, head of, 86.

Megrez, 41.

Menkalina, 112.

Menkar, 114.

Merak, 41.

Mercury, 122.

Meridian, 13.

Mesarthim, 84.

Milky Way, 17, 18, 27, 56, 64, 72.

Mintaka, 102.

Mira, 114.

Mirach, 81.

Mirrors, sky views by, 19.

Mitchel, Gen. O. M., 55.

Mizar, 41, 47.

Morning of the year, 21.

Moses and the Brazen Serpent, 57.

Mukdim-al Kitaf, 31.

Murzim, 100.

Mut, temple of, 46.

Mystery in the sky, 58.

8 M., 56.

Names of stars and travellers, 76.

Nautical Almanac, 122.

Nebulæ, in Andromeda, 82.

in Canes Venatici, 69.

in Lyra, 63.

in Ophiuchus, 58.

in Orion, 103.

in Sagittarius, 56.

in Virgo, 32.

New star of 1901, 90.

Northern Cross, 63.

Northern Crown, 66.

North star, 46, 116.

November meteors, 37.

Nu Scorpii, 55.

Omicron Ceti, 114.

Ophiuchus, 57.

Orion, 101 *et seq.*

Orion group of constellations, 40.

Pearl, the, 66.

Pegasus, 80.

Perseus, 85.

Phæd, 41.

Phæton, 84.

Phosphorus, 123.

Pi Orionis, 105.

Pisces, 85, 115.

Piscis Austrinus, 75.

Planets, the, 118;

apparent swallowing by sun, 125.

Plato, quoted, 39.

Pleiades, 109 *et seq.*

Pointers, the, 42.

Polaris, 46, 116.

Pole-stars, succession of, 48.

Pollux, 113.

Porrima, 32.

Præsepe, 38.

Precession of equinoxes, 47, 84.

Procyon, 113 *et seq.*

Pulcherrima, 69.

Ras Algethi, 59 *et seq.*

Ras Alhague, 57.

Reflection, sky seen by, 19;

supposed visibility of Jupiter's moons by, 124.

Regulus, 35 *et seq.*

Revelation of the stars, 10.

Revolution of the pole, 48.

Revolutions of the heavens, 16.

Rho Ophiuchi, 58.

Richter, Jean Paul, quoted, 91.

Rigel, 104.

Right Ascension, 13.

Rising stars, attraction of, 29.

Royal family of sky, 75.

Royal stars, 35.

Ruchbah, 88.

Sadachbia, 78.

Sadalmelik, 77.

Sadalsuud, 78.

Sagitta, 79.

Sagittarius, 56.

St. Paul and the viper, 57.

Saiph, 105.

Saturn, 122, 124.

Scheat, 81.

Schedar, 88.

Scorpio, 55.

Serpens, 57.

Set, 122.

Shakespeare, quoted, 49, 80.

Sheratan, 84.

Shield of Orion, 105.

Sickle, the, 37.

Sigma Tauri, 109.

Sirius, 94 *et seq.*

Smyth, Admiral, quoted, 32.

Sobieski's Shield, 57, 65.

Solstices, 13.

Sophocles, quoted, 67.

Southern Cross, R. H. Dana on, 77.

Southern Fish, 75.

Spica, 29 *et seq.*

Star colors, 44, 83, 89, 90.

Star magnitudes, 98.

Streaming of stars, 81, 106.

Struve invents star color, 106.

Summer Solstice, 50.

Sword-hand of Perseus, 86.

Sword of Orion, 103.

Tent, the, 33.

Theta Orionis, 103.

Theta Tauri, 109.

Three Guides, the, 88.

Thuban, 45, 48.

Toorus, 107.

Tropic of Capricorn, 79;

of Cancer, 40.

Tycho's star, 90.

Ursa Major, 40.

Ursa Minor, 46.

Uru-anna, 101.

Vega, 43, 61 *et seq.*

Venus, 120, 123.

Vernal Equinox, 21.

Vindemiatrix, 31.

Virgin, origin of name, 30.

Virgo, 29 *et seq.*

Vulpecula et Anser, 80.

"W," the letter, 87.

Wesen, 100.

Whirlpool nebula, 69.

Whitney, Prof., quoted, 78.

Winter heavens, glories of, 93.

Winter Solstice, 93.

"XM" class of stars, 31, 64, 104.

Xavier de Maistre, quoted, 89.

Year, various beginnings of, 25 *et seq.*

Zeta Ursæ Majoris, 40.

Aquarii, 78.

Canis Majoris, 100.

Herculis, 60.

Orionis, 102, 106.

Zodiac, 84.

Zubeneschemali, 70.

Zubenelgenubi, 70.

THE END